THE EXACT SCIENCES
IN ANTIQUITY

THE EXACT SCIENCES
IN ANTIQUITY

BY

O. NEUGEBAUER

Second Edition

DOVER PUBLICATIONS, INC.
NEW YORK

Published in Canada by General Publishing Company, Ltd., 30 Lesmill Road, Don Mills, Toronto, Ontario.
Published in the United Kingdom by Constable and Company, Ltd., 10 Orange Street, London WC 2.

This Dover edition, first published in 1969, is an unabridged and slightly corrected republication of the second edition, published in 1957 by Brown University Press.

Standard Book Number: 486-22332-9
Library of Congress Catalog Card Number: 69-20421

Manufactured in the United States of America
Dover Publications, Inc.
180 Varick Street
New York, N. Y. 10014

To

RICHARD COURANT

in Friendship and Gratitude

FROM THE PREFACE TO THE FIRST EDITION

The first series of Cornell University's "Messenger Lectures on the evolution of civilization" was given by James Henry Breasted, eminent Egyptologist and founder of the Oriental Institute of the University of Chicago. Few scholars have contributed so much to our understanding of ancient civilizations and have attracted the interest of scholars and laymen alike to the study of the ancient Near East. I personally feel a great debt of gratitude towards Breasted whose "History of Egypt" was my first stimulus towards the study of ancient oriental civilizations, a field of research which has occupied me ever since and about whose role in the history of science I shall report in the following pages. That I was able to follow this road from the early days of my graduate study in Göttingen is due to the never failing encouragement and support of R. Courant. But more than that I owe him the experience of being introduced to modern mathematics and physics as a part of intellectual endeavor, never isolated from each other nor from any other field of our civilization. In dedicating these lectures to him I only acknowledge publicly a debt which has profoundly influenced my own development.

The following chapters follow closely the arrangement of six lectures which I delivered at Cornell University in the fall of 1949. I fully realize that this form of presentation forced me into many statements which actually should be qualified by many conditions and question marks. I also realize that the following pages will give ample opportunities to quote statements and to utilize them in a sense which I did not imply or did not foresee. And I have no doubts that many a conclusion will have to be modified and corrected. I am exceedingly sceptical of any attempt to reach a "synthesis"—whatever this term may mean—and I am convinced that specialization is the only basis of sound knowledge. Nevertheless I have enjoyed the possibility of being compelled for once to abandon all learned apparatus and to pretend to know when actually I am guessing. This does not imply that I have ignored facts. Indeed, I have consistently tried to keep as close as possible to the source material. Only in its selection, in its arrangement, and in its coherent interpretation have I permitted myself much greater freedom than is usual in technical publications. And in order to counteract somewhat the impression of security which easily emerges from general discussions I have often inserted methodological remarks to remind the reader of the exceedingly slim basis on which, of necessity, is built any discussion of historical developments from which we are separated by many centuries. The common belief that we gain "historical perspective" with increasing distance seems to me utterly to misrepresent the actual situation. What we gain is merely confidence in generalizations which we would never dare make if we had access to the real wealth of contemporary evidence.

The title "The Exact Sciences in Antiquity" is not meant to suggest an exhaustive discussion of this vast subject. What I tried to present is a survey of the historical interrelationship between mathematics and astronomy in ancient civilizations, not a history of these disciplines in chronological arrangement. Since the works of Sir Thomas L. Heath provide an excellent guide for Greek mathematics, I see no need to summarize their contents in a series of lectures. For Greek astronomy no similar presentation exists, but its highly technical character makes it impossible to discuss any details in the present book. Consequently, the main emphasis is laid on mathematics and astronomy in Babylonia and Egypt in their relationship to Hellenistic science.

In the notes which follow the single chapters I have added some technical details which seem to me relevant to further study. I have also quoted several works which lead far away from the direct path of the present approach because I feel that this book will have best reached its goal if it convinces the reader that he finds here only one of many ways of approach to a subject which is much too rich to be exhausted in six chapters.

Instead of attempting completeness I have tried to convey to the reader some of the fascination which lies in active work on historical problems. I wished to confront him with one of the ever-changing pictures which one forms as a kind of guiding principle for future research.

PREFACE TO THE SECOND EDITION

*Tu deviens responsable
pour toujours de ce que
tu as apprivoisé*
SAINT-EXUPÉRY

When preparing this second edition, I was helped again by my friends and colleagues, particularly by A. Sachs, but I alone am responsible for any statements which might be incorrect or might become untenable in the light of further research.

I am very thankful to Brown University and in particular to its librarian Mr. D. A. Jonah for having made possible the publications of this book. Mr. Torkil Olsen in Copenhagen was very helpful in arranging for the printing which was completed with traditional craftsmanship in Odense, Denmark.

In order to keep this book up-to-date, many additions referring to recently obtained results have been made. Large sections on Egyptian astronomy and on Babylonian planetary theory have been rewritten. Two appendices are entirely new, one on Greek Mathematics, the other on the Ptolemaic system and its Copernican modification.

I hope I have avoided, in spite of these amplifications, converting my lectures into a textbook.

O. N.

TABLE OF CONTENTS

LIST OF PLATES

(Plates 1–14 appear following p. 240)

"And when he reaches early adolescence he must become possessed with an ardent love for truth, like one inspired, neither day nor night may he cease to urge and strain himself in order to learn thoroughly all that has been said by the most illustrious of the Ancients. And when he has learnt this, then for a prolonged period he must test and prove it, observe what part is in agreement, and what in disagreement with obvious facts; thus he will choose this and turn away from that. To such a person my hope has been that my treatise would prove of the very greatest assistance. Still, such people may be expected to be quite few in number, while, as for the others, this book will be as superfluous to them as a tale told to an ass."

GALEN, *On the natural faculties*, III, 10.

[Translation by ARTHUR JOHN BROCK, M. D.
The Loeb Classical Library p. 279/281.]

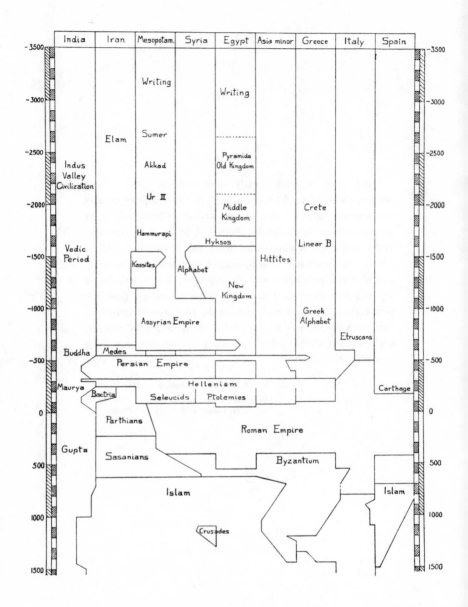

THE EXACT SCIENCES
IN ANTIQUITY

INTRODUCTION

The investigation of the transmission of mathematics and astronomy is one of the most powerful tools for the establishment of relations between different civilizations. Stylistic motives, religious or philosophical doctrines may be developed independently or can travel great distances through a slow and very indirect process of diffusion. Complicated astronomical methods, however, involving the use of accurate numerical constants, require for their transmission the direct use of scientific treatises and will often give us very accurate information about the time and circumstances of contact. It will also give us the possibility of exactly evaluating the contributions or modifications which must be credited to the new user of a foreign method. In short the inherent accuracy of the mathematical sciences will penetrate to some extent into purely historical problems. But above and beyond the usefulness of the history of the exact sciences for the history of civilization in general, it is the interest in the role of accurate knowledge in human thought that motivates the following studies.

The center of "ancient science" lies in the "Hellenistic" period, i. e., in the period following Alexander's conquest of the ancient sites of oriental civilizations (Frontispiece). In this melting pot of "Hellenism" a form of science was developed which later spread over an area reaching from India to Western Europe and which was dominant until the creation of modern science in the time of Newton. On the other hand the Hellenistic civilization had its roots in the oriental civilizations which flourished about equally long before Hellenism as its direct influence was felt afterwards. The origin and transmission of Hellenistic science is therefore the central problem of our whole discussion.

I restrict my subject to the exact sciences simply because I feel totally incompetent to deal with subjects like medicine or the natural sciences, though much important information could be obtained for our problem from an investigation of these fields. Medicine and astronomy, for example, are closely related in the Greek medical schools; similarly, medieval medicine was deeply affected by Hellenistic astrology. The sciences of drugs, plants, stones, and even the animal kingdom show many points of contact with astronomical or astrological doctrines; our use of the name of Mercury for a substance and for a planet is a still living witness of this. Nor did medieval or Renaissance artists pride themselves on being ignorant of the sciences. The sculptures of Gothic cathedrals and the paintings and miniatures of the Middle Ages are full of astronomical or astrological references which were significant to the contemporary man. Thus, it is a quite artificial restriction which we impose upon the following discussions in limiting ourselves to exact mathematics and mathematical astronomy.

And even within these narrow limits it was necessary to lay undue weight on the part of mathematics as compared with astronomy. The basic mathematical concepts are simple and much more familiar to the modern reader than the corresponding astronomical facts and their ancient presentation, which often will be rather strange even to a professional modern astronomer. All I could hope to do within the given framework was to remind the reader on many occasions of the paramount role which astronomy played in the history of science. I do not hesitate to assert that I consider astronomy as the most important force in the development of science since its origin sometime around 500 B.C. to the days of Laplace, Lagrange, and Gauss. And I hasten to say that the history of the origin of astronomy is one of the most fragmentary chapters in the history of science, however great our gaps may be for other periods and other problems. Consequently I am convinced that the history of mathematical astronomy is one of the most promising fields of historical research. I hope that this will become evident, at least to a certain extent, to any reader of the following chapters.

CHAPTER I

Numbers.

1. When in 1416 Jean de France, Duc de Berry, died, the work on his "Book of the Hours" was suspended. The brothers Limbourg, who were entrusted with the illuminations of this book, left the court, never to complete what is now considered one of the most magnificent of late medieval manuscripts which have come down to us.

A "Book of Hours" is a prayer book which is based on the religious calendar of saints and festivals throughout the year. Consequently we find in the book of the Duke of Berry twelve folios, representing each one of the months. As an example we may consider the illustration for the month of September. As the work of the season the vintage is shown in the foreground (Plate 1). In the background we see the Château de Saumur, depicted with the greatest accuracy of architectural detail. For us, however, it is the semicircular field on top of the picture, where we find numbers and astronomical symbols, which will give us some impression of the scientific background of this calendar. Already a superficial discussion of these representations will demonstrate close relations between the astronomy of the late Middle Ages and antiquity. This is indeed only one specific example of a much more general phenomenon. For the history of mathematics and astronomy the traditional division of political history into Antiquity and Middle Ages is of no significance. In mathematical astronomy ancient methods prevailed until Newton and his contemporaries opened a fundamentally new age by the introduction of dynamics into the discussion of astronomical phenomena. One can perfectly well understand the "Principia" without much knowledge of earlier astronomy but one cannot read a single chapter in Copernicus or Kepler without a thorough knowledge of Ptolemy's "Almagest". Up to Newton all astronomy consists in modifications, however ingenious, of Hellenistic astronomy. In mathematics the situation is not much different though the line of demarcation between "ancient" and "modern" is less sharply drawn. But also here the viewpoint could well be defended that all "modern" mathematics begins with the creation of analysis, thus again with Newton and his contemporaries.

2. We shall not worry, however, about historical doctrines. Merely as an illustration for the continuation of ancient traditions we shall briefly analyze the calendars of the Book of the Hours of the Duke of Berry. Both the outermost and the inner ring contain numbers, the inner ring from 1 to 30, the outer ring from 17 to 30 and from 1 to 15. The appearance of these numerals is not quite the one familiar to us today—cf., e. g., the 14 and 15 at the right end of the outer circle—but everyone will easily decipher their values and recognize that they are the familiar "Hindu-Arabic" numerals which penetrated into Europe, beginning in the 12th century, from the Islamic world. They superseded more and more the Roman numerals of which we find also two representatives in our picture. The right half of the outer rim shows the inscription *initium libre gradus XV* "beginning of Libra 15 degrees" and in the inner rim we read *primationes lune mensis septembris dies XXX* "the primationes of the moon of the month of September, 30 days". Implicitly we have here a representation of a third method of expression of numerals, namely, by number words. September, and similarly October, November, December, are denominations of the months as 7, 8, 9, 10 respectively—thus reflecting a period of the Roman calendar when the end of the year fell two months later than in our present calendar.

3. These three types of numerical expression can be found in many examples all over the world. The writing of numbers by simple words without the use of any symbols whatsoever is very common indeed. A variant of it is the method found in Greek inscriptions which use abbreviations like Π for $\Pi ENTE$ "five" or $\varDelta$ for $\varDelta EKA$ "ten". One calls this the "acrophonic" principle, where the first letter stands for the whole word.

The Roman system is perhaps the most widespread method, historically speaking. The smallest numbers are simple repetitions of 1. This holds even for our present number symbols where 2 and 3 originated from $=$ and $\equiv$ by connecting lines in cursive writing. The same system prevailed in Egypt, in Mesopotamia, or, for the smallest units, in the Greek inscriptions just mentioned. Roman V is probably half of the symbol X as D ($= 500$) is half of $\Phi = 1000$. This latter symbol was only later conveniently interpreted as M for "mille" thousand. Similar individual symbols for 10, 100, 1000 are found in Egypt. Their repetition and combination readily yields the intermediate numbers. Arrangement may play a role, as in IV $= 5 - 1$ in contrast to VI $= 5 + 1$. As an explicit case of subtractive writing may be mentioned an Old-Babylonian form for 19. In this period "one" would be Γ; "ten," $\langle$; thus $21 = $ ⟨𝍥 but ⟨𝍥$\Gamma = 20 - 1 = 19$. Here the sign Γ LAL "subtract" is written between 20 and 1. Later, 19 would be written only ⟨⟨𝍦 $= 10 + 3 + 3 + 3$ from which a final cursive form ⟨⟨𝍦 originated in the Seleucid period.

Fundamentally different from all these methods is the "place value notation" of our present system, where neither 12 nor 21 represents $1 + 2$ or $2 + 1$ but 1 times ten plus 2, and 2 times ten plus 1 respectively. Here the position of a number symbol determines its value and consequently a limited number of symbols suffices to express numbers, however large, without the need for repetitions or creation of new higher symbols. The invention of this place value notation is undoubtedly one of the most fertile inventions of humanity. It can be properly compared with the invention of the alphabet as contrasted to the use of thousands of picture-signs intended to convey a direct representation of the concept in question.

4. Before returning to the history of number symbols we shall draw some additional information from the calendar of the Book of Hours. The wide middle zone shows the pictures of "Virgo" and the scales of "Libra", headed by the inscriptions *finis graduum virginis* "end of the degrees of Virgo" and the already quoted "beginning of Libra 15 degrees." Virgo and Libra are signs of the zodiac, i. e. sections of 30 degrees each in the yearly path of the sun among the fixed stars as seen from the earth. Consequently our picture indicates that the sun travels during September from the 17th degree of Virgo to the 15th degree of Libra,

or a total of 29 degrees, as can be counted directly by tallying
the spaces on the outer rim. Because September has 30 days the
sun covers in one day $\dfrac{29}{30} = \dfrac{58}{60}$ degrees or 58 minutes of arc per
day. This corresponds very well to the facts. Because it takes the
sun slightly more than 365 days to travel the 360 degrees of the
whole zodiac, the average daily travel must be slightly less than
one degree per day. If we repeat our computation for all the
12 folios of our calendar we find, however, a faster movement
of 1° per day for November, December, and January. This is
counterbalanced by a slower movement of about 56 minutes in
the months from March to July. This again reflects facts correctly.
The sun moves fastest in Winter, slowest in Summer; and we
shall see that this phenomenon was accurately taken into consideration both in Greek and in Babylonian astronomy of the
Hellenistic period. One calls this irregularity of the solar motion
its "anomaly". It is certainly not to be expected a priori to find
this concept carefully represented in a prayer book of the early
15th century.

5. An additional numerical notation occurs in the inner ring
of the calendar. Here we find associated with symbols of the
moon the following letters: $b\ k\ s\ g\ f\ d\ m\ a\ i$ etc. If we assign
to these letters numbers according to their position in the alphabet
we obtain:

2 10 18 7 6 4 12 1 9 17 6 14 3 11 19 8 16 5 13

These numbers are obviously connected by the following simple
law: always add 8 to the preceding number in order to get the
next number; in case the total exceeds 19, subtract 19. Thus

$2 + 8 = 10 \qquad 10 + 8 = 18 \qquad 18 + 8 = 26; \quad 26 - 19 = 7.$

The next number should be $7 + 8 = 15 = p$ followed by
$15 + 8 = 23; \quad 23 - 19 = 4 = d$. The text, however, has f
followed by d. Hence we must correct f into p, and this correction
is confirmed by the calendars for the other months where we
always find the arrangement $g\ p\ d$. The remaining part of the list
is correct. In the last place we have $5 + 8 = 13$ to be followed by
$21 - 19 = 2$ which is the first number of our list. Thus the list
repeats itself after 19 steps.

The question as to the significance of the number 19 leads us directly back to the 5th century B.C. when a cyclic scheme of intercalations was actually introduced in the Babylonian calendar and unsuccessfully proposed in Athens by Meton, who was, however, honored by his contemporaries with a statue and by modern scholars with the attachment of his name to the cycle.

The basis of this cycle can be explained very simply as follows. The time between two consecutive conjunctions of sun and moon is about $29\frac{1}{2}$ days. This interval is called one "lunation." A lunar month is therefore either 29 or 30 days long. Consequently 12 lunar months amount to 354 days or about 11 days less than one solar year. After three years a deficiency of about 33 days has accumulated, making it necessary to add a 13th month to one of the three lunar years in order to bring the beginning of the lunar year roughly back to the beginning of the solar year. More accurate recording of the beginnings of lunar months and the beginnings of solar years shows that 19 solar years contain 235 lunar months, i. e., 12 ordinary lunar years of 12 months each and 7 intercalary lunar years of 13 months each. This 19-year or Metonic cycle is quite accurate; only after 310 Julian years do the cyclically computed mean new moons fall one day earlier than they should. This simple cyclical computation not only formed the basis of the calendar of the Seleucid empire in antiquity but is similarly the foundation of the Jewish and Christian religious calendar, especially so far as Easter is concerned. The same cycle appears, though in a slight disguise, in the luni-solar computations of two of the earliest astronomical works of India, the Romaka- and the Paulisa-Siddhānta (about fifth century A.D.), whose Western origin is apparent from their names and confirmed by many details.

6. By means of this cycle the Middle Ages solved the problem of establishing the dates of the new moons, at least for purposes of the religious calendar, though the actual facts might differ by several days. The *"primationes lunae"* or new moons in our Book of the Hours are determined as follows: As the first year, "a," of the cycle a year is chosen when the new moon fell on January 19 (cf. Fig. 1). From now on we operate with alternating lunations of 30 or 29 days respectively, with occasional additions of one day such that two 30-day months follow one another. In

this way one obtains February 18 for the next new moon, (30 days after January 19), then March 19 (29 days) after February 18, etc. Continuing this process[1]) we reach September 13 as a new-moon date for "year a" and indeed the letter "a" is given at this date below a little crescent in our calendar miniature for

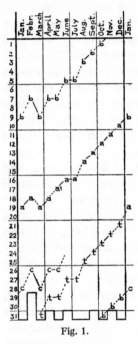

September. Continuing with alternating 29 and 30 day lunations we reach January 9 of the second year, called "b." For September we find "b" marked at day 2; for October one finds October 1 and October 31 for year "b," etc. This procedure leads eventually to an arrangement of letters, representing the numbers from $a = 1$ to $t = 19$, exactly in the form which we see in the special case of September. The scheme ends where it began, with January 19, if we make the two last lunations 29 days long. This final exception to the rule of alternation was called *saltus lunae*, the "jump of the moon".

In order to know which date is supposed to be a new moon one need only know which number the present year has in the 19-year cycle. This number is called the "golden number" because, as a scholar of the 13th century expressed it, "this number excels all other lunar ratios as gold excels all other metals." In the twelfth century this very primitive method was considered by scholars in Western Europe as a miracle of accuracy, though incomparably better results had been reached by Babylonian and Greek methods since the fourth century B.C. and though these methods were ably handled by contemporary Islamic and Jewish astronomers.

7. Scientific progress can perhaps be best measured by the number of previously separated facts which become understandable under a new common viewpoint. By this standard the recession from a lunar ephemeris of the third century B.C. to a

Fig. 1.

---

[1]) One pair of 30-day lunations was inserted between March and July. In Fig. 1 30-day intervals are indicated by solid lines, 29-day intervals by dotted lines.

lunar calendar 1700 years later is rather drastic, even disregarding all astronomy and looking at simple arithmetic only. Instead of one uniform numerical system, several computing devices are used side by side. This even goes so far that for the handling of the above-mentioned lunar letters special tricks were invented to establish their sequence by means of the segments of the fingers to which these letters were assigned. Thus one has come back to a wide-spread technique of "computing on the fingers". These methods form a substratum of primitive mathematical lore which has been found in the most different ancient civilizations, as well as among nations of the Near and Far East. Probably its earliest occurrence is in Egypt in a passage of the "Book of the Dead" which in turn is based on a spell of the "Pyramid Texts." In the "Spell for Obtaining a Ferry-Boat" the deceased king tries to convince the ferryman to let him cross a canal of the nether world over to the Eastern side. To this the ferryman objects with the words: "This august god (on the other side) will say, 'Did you bring me a man who cannot number his fingers?'." But the deceased king is a great "magician" and is able to recite a rhyme which numbers his ten fingers and thus satisfies the requirements of the ferryman. It seems obvious to me that we are here reaching back into a level of civilization where counting on the fingers was considered a difficult bit of knowledge of magical significance, similar to being able to know and to write the name of a god. This relation between numbers (and number words) and magic remained alive throughout the ages and is visible in Pythagorean and Platonic philosophy, the Kabbala, and various other forms of religious mysticism.

8. We return once more to the diverse methods of writing numbers. Four different types of writing can be illustrated on the calendar of the Book of Hours: the place value notation still in use today; the Roman numerals operating with individual symbols for the different groups of units; complete number words; and finally alphabetic numerals. From Greek inscriptions we have added a fifth method, the "acrophonic" writing which consists, however, only in an abbreviation of number words. We shall now discuss other variants of Greek numerical notation.

The first is strictly alphabetical and is found on Athenian coins of the second century B.C. On these coins, of the so-called

"New Style", the months of issue are denoted by the letters *A*
to *M* representing the numbers 1 to 12 for an ordinary year,
adding *N* = 13 for a leap year of the Athenian lunar calendar.
The same principle is followed in Ptolemaic coins from Egypt
where the numbers *AA BB ΓΓ* etc. occur, obviously indicating
25, 26, 27 etc. after the first 24 letters from *A* to *Ω* were exhausted. It is clear how one in prin-
ciple could continue this system.

9. Much more important, however, is another modification of
the alphabetic numeration which
is extensively used in Greek mathematics and astronomy and
also in economic and literary
documents, e.g. in Greek papyri.
Though this system of Greek
numerals is often described in
books on the history of mathe-
matic and elsewhere, I shall
sketch the way one might be able
to decipher this system in any
sufficiently elaborate mathematical or astronomical text. This
can at the same time serve as
an illustration of how one proceeds in similar cases with less
well known context.

ια'. Κανόνιον τῶν ἐν κύκλῳ εὐθειῶν.

περιφε-ρειῶν	εὐθειῶν			ἑξηκοστῶν			
ζ'	ο	λα	κε	ο	α	β	ν
α	α	β	ν	ο	α	β	ν
αζ'	α	λδ	ιε	ο	α	β	ν
β	β	ε	μ	ο	α	β	ν
βζ'	β	λζ	δ	ο	α	β	μη
γ	γ	η	κη	ο	α	β	μη
γζ'	γ	λθ	νβ	ο	α	β	μη
δ	δ	ια	ιϛ	ο	α	β	μζ
δζ'	δ	μβ	μ	ο	α	β	μζ
ε	ε	ιδ	δ	ο	α	β	μϛ
εζ'	ε	με	κϛ	ο	α	β	μϛ
ϛ	ϛ	ιϛ	μθ	ο	α	β	μδ
ϛζ'	ϛ	μη	ια	ο	α	β	μγ
ζ	ζ	ιθ	λγ	ο	α	β	μβ
ζζ'	ζ	ν	νδ	ο	α	β	μα
η	η	κβ	ιε	ο	α	β	μ
ηζ'	η	νγ	λε	ο	α	β	λθ
θ	θ	κδ	να	ο	α	β	λη
θζ'	θ	νϛ	ιγ	ο	α	β	λζ
ι	ι	κϛ	λβ	ο	α	β	λϛ
ιζ'	ι	νη	μθ	ο	α	β	λγ
ια	ια	λ	ε	ο	α	β	λβ
ιαζ'	ιβ	α	κα	ο	α	β	λ
ιβ	ιβ	λβ	λϛ	ο	α	β	κη

Fig. 2.

I take as our example a table from Ptolemy's "Almagest"
(Fig. 2). The heading says "Table of straight lines in the circle,"
i. e. table of chords. The first column is described as "arcs."
In this column we find in every second line the familiar Greek
letters in the arrangement of the alphabet. We are obviously
dealing with numbers; thus we make the simplest assumption
$\alpha = 1$ $\beta = 2$ $\gamma = 3$ $\delta = 4$ $\varepsilon = 5$. Thereafter one should
expect $\zeta = 6$ but ζ appears only one step later and we are forced
to read the intermediate symbol ς as 6. Thereafter we obtain
again a regular sequence $\zeta = 7$ $\eta = 8$ $\theta = 9$ $\iota = 10$. Following
the alphabetic order one might expect $\varkappa = 11$ $\lambda = 12$ etc.
Actually, however, we find $\iota\alpha = 11$ $\iota\beta = 12$ etc., in other words,

combined symbols $10 + 1$, $10 + 2$, etc. This is readily confirmed by the continuation of our table (not reproduced here) where at the proper place $\iota\theta = 19$ is followed by $\varkappa = 20$ $\varkappa\alpha = 21$ etc. Continuing in this fashion one will meet once more a disturbance of the standard alphabet when after $\pi = 80$ a strange sign φ signifies 90. Then follows $\varrho = 100$ $\sigma = 200$ $\tau = 300$ until $\omega = 800$, followed again by a special symbol $\mathbf{\tau}$ (or $\mathbf{\Lambda}$ or $\mathbf{\varphi}$) $= 900$.

Though the three symbols ς φ and $\mathbf{\varphi}$ are not members of the classical Greek alphabet they are well known to the historian as remnants of the earliest form of the Greek alphabet which still shows these three letters in actual use. Consequently the alphabetic numerals were invented when the Greek alphabet had not yet eliminated these three sounds which it took over with the rest of the alphabet from the Phoenicians. Considerations of this type allow us to date the origin of the Greek alphabetic number system to about the 8th century B.C. and to localize its invention with great probability at the city of Miletus in Asia Minor.

Returning to our table we have still omitted every second line. It is obvious, however, that one would guess that the sign $\angle'$ represents $\frac{1}{2}$ because we then can read

$$\frac{1}{2} \quad 1 \quad 1\frac{1}{2} \quad 2 \quad 2\frac{1}{2} \quad 3 \quad 3\frac{1}{2} \quad \text{etc.}$$

We can confirm this hypothesis immediately by means of the second column. This column contains three subcolumns which we already can transcribe by means of our previous decipherment with the exception of the new symbol $\circ$ in the very first place. Calling this symbol x we read

x	31	25
1	2	50
1	34	15
2	5	40
2	37	4
3	8	28

etc. The structure of these three columns of numbers is obvious. In the second and third column we observe alternatingly smaller and larger numbers whereas the numbers in the first column either remain unchanged or increase by one. This last observation compels us to assign to x the value "zero." Consequently it is plausible to consider the numbers in the second and third column as fractions and to assume that the numbers as a whole increase from 0 to 1, 2, etc. Indeed, if we look at the last numbers we find that they increase from 25 to 50, then fall down to 15 but increase again by 25 to the next 40. We would have a constant increase by 25 if we had the sequence 25 50 60 + 15 60 + 40 or if 60 units of the last place would amount to 1 of the preceding place. This is easily tested in the preceding column. The numbers 31 2 34 5 37 show again almost constant increase if we take a total of 60 as one higher unit:

$$31 \qquad 31 + 31 = 60 + 2 \qquad\qquad 62 + 32 = 60 + 34$$
$$\qquad 60 + 34 + 31 = 120 + 5 \qquad 120 + 5 + 32 = 120 + 37$$

etc.

The increase is either 31 or 32 and it is 32 when and only when 60 units of the third column have accumulated. And whenever 60 units of the second column have accumulated, the number in the first column increases by one. Thus we have a system of numbers which behave exactly like degrees, minutes, and seconds, or like hours, minutes, and seconds; the fractions are sixtieths of the next higher unit. We call such fractions "sexagesimal fractions" and write numbers of this type in the following form:

$$0,31,25$$
$$1,2,50$$
$$1,34,15$$
$$2,5,40$$

We can say that these numbers show a constant difference 0,31,25. Later in our table the differences become smaller and smaller, but this is exactly what one should expect. If the first column indicates arcs increasing by $\frac{1}{2}$ degree, as is indicated by the numbers already known, then we must expect that the chords do not grow simply proportionately with the arcs, though this might hold for very small angles at the beginning of the table. But what are the units used in our table? That the first column indicates degrees is obvious from the fact that the table ends with 180, i. e., with the straight angle. The chord to 180° must be the diameter; the table gives for this entry the value ϱϰ ο ο = 120,0,0. Thus the radius is 60. This is confirmed by the chord 60 for 60°, as is correct for the equilateral triangle where chord = radius. We do not need to discuss in detail the third column, called "sixtieths," of the table of chords. We know already that the chord of 1° is 1,2,50. Hence the chord for 0;1° (or for 1 minute of arc[1])) will be 0;1,2,50 as given in the third column. In general the third column gives the coefficients of interpolation for single minutes, as is easily confirmed from the section reproduced on p. 10.

10. Our example of a Greek numerical table familiarized us with several interesting features of the most important type of Greek numerals. The borrowing of the Greek alphabet from the Phoenicians explained the symbols for 6, 90, and 900. We found a special sign for $\frac{1}{2}$, a phenomenon which could be amplified from papyrus documents and other sources. We found a special sign for zero, used exactly as our zero. And finally we have seen the sexagesimal system in full use, both in the familiar division of the circumference of the circle into 360 "degrees" of 60 minutes or 3600 seconds each, and in the division of the radius into units of consecutive sixtieths.

These features are not restricted to an isolated case like the table of chords which we quoted. All Greek astronomical works, containing hundreds of extensive numerical tables, are based on exactly the same procedure. According to the prevailing doctrine that Greek mathematics is essentially geometry, the historians of mathematics have badly neglected the enormous amount of numerical computations which are readily accessible in works like Ptolemy's "Almagest" or Theon's "Handy Tables." But long before these classics were written, Greek astronomical papyri were covered with computations. While Ptolemy or Theon are today preserved only in Byzantine manuscripts, we do have papyri from the Ptolemaic period[2]) onwards. In these papyri we can find, e. g., the zero sign as it was actually written. An example is a papyrus written in the second century A.D. (cf. Pl. 2 and the transcription on p. 163). Near the end of the last line

---

[1]) I apply here a notation which will be used throughout in the subsequent pages. A semicolon separates integers from fractions, while all other sexagesimal places are separated from one another by a comma. Thus 1,1 means 61 but 1;1 = $1 + \frac{1}{60}$. Neither comma nor semicolon has any counterpart in the actual texts.

[2]) "Ptolemaic period" refers to the dynasty of the Ptolemies who ruled over Egypt during the last three centuries before our era. Ptolemy the astronomer, about 150 A.D., has nothing to do with this dynasty.

preceding the empty line in the upper part of the papyrus, one finds representing zero a sign which looks like ⚬⚬. In other astronomical papyri are found similar symbols varying from forms like ⚬ or ⚬ to ⚬. In the form ō and related variants this zero symbol is found until the latest periods in Arabic geographical and astronomical manuscripts where numbers were written in the alphabetic notation. Only in Byzantine manuscripts do I know of the bare *o*-like shape which is usually considered as the first letter of Greek $ουδεν$ "nothing." The papyri do not support this explanation (which is in itself very implausible since omicron already represented a numerical value, namely 70) but suggest an abitrarily invented symbol intended to indicate an empty place. This would correspond exactly to the Babylonian zero symbol which is also not a letter or a syllable but a mere separation mark.

    11. In order to make this remark fully understandable, I have to explain briefly a main point in the chronology of "Babylonian" mathematical and astronomical source material. The texts of which I speak are clay tablets, generally about the size of a hand, inscribed with signs which were pressed into the surface of the once soft clay by means of a sharpened stylus. This script is called "cuneiform," i. e. wedge-shaped, because the individual impressions have a deeper "head" and a finer line at the end, thus resembling a wedge. Cuneiform tablets with mathematical contents are known to us mostly from the so-called "Old-Babylonian" period, about 1600 B.C. (cf. Pl. 3). No astronomical texts of any scientific significance exist from this period, while the mathematical texts show the highest level ever attained in Babylonia.

    The second period from which we have a larger number of texts is the latest period of Babylonian history, when Mesopotamia had become a part of the empire of Alexander's successors, the "Seleucids." This period, from about 300 B.C. to the beginning of our era, has furnished us with a great number of astronomical texts of a most remarkable mathematical character, fully comparable to the astronomy of the Almagest. Mathematical texts from this period are scarce, but they suffice nevertheless to demonstrate that the knowledge of Old-Babylonian mathematics had not been lost during the intervening 1300 years for which texts are lacking.

Thus it is essential to remember that we are dealing with mathematical texts from two periods, "Old-Babylonian" from about 1800 to 1600, and "Seleucid" from 300 to 0, whereas astronomical texts belong only to the second period.

12. The development of the numerical notations in Mesopotamia took as many centuries as the development of writing from a crude picture script to a well defined system of complicated signs. We shall for the moment deal only with the final product as it appears in the mathematical texts of the Old-Babylonian period. And we shall again use the most direct approach by deciphering an actual text.

Plate 4,a shows a tablet whose size is about 3⅛ by 2 inches (and about ¾ of an inch thick). In the middle of the text is visible a column of signs which obviously represent numbers in ascending order. The tablet is not quite cleaned from incrustation of salt or dirt but it is clear that the signs look about as follows:

𒁹 𒈫 𒐈 𒐉 𒐊 𒐋 𒐌 𒐍 𒐎 𒌋 𒌋𒁹 𒌋𒈫 𒌋𒐈

Counting of the vertical wedges leads directly to the readings 1, 2, 3, etc. up to 9. Then follows 𒌋 which must be 10, and consequently we can also read the remaining signs as 11, 12, 13. Using this exceedingly plausible hypothesis, we should also be able to read the right-hand column of signs. The first five look as follows

𒌋 𒌋𒌋 𒌍 𒐏 𒐐

Obviously we must read these signs as 10, 20, 30, 40, 50 if the first sign represents 10 as we have established in our first list. But what follows is

𒁹   𒐕   𒐖   *   *   *   𒈫   𒐗

which we transcribe consistently as

$$1 \quad 1{,}10 \quad 1{,}20 \quad * \quad * \quad * \quad 2 \quad 2{,}10$$

each * indicating a broken line. These signs continue the previous ones if we interpret the first "1" as 60 and then read 1,10 as

$60 + 10 = 70$ and $1,20$ as $60 + 20 = 80$. The broken lines should contain 90, 100, and 110. The next sign "2" should be 120, in excellent agreement with our interpretation of "1" as 60, while the last sign $2,10$ must be $120 + 10 = 130$. Thus we have obtained all multiples of 10 from 10 to 130, line by line, corresponding to the numbers 1 to 13. In other words, our table is a multiplication table for 10, which we now can transcribe as follows:

1	10
2	20
3	30
4	40
5	50
6	1
7	1,10
8	1,20
9	1,30
10	1,40
11	1,50
12	2
13	2,10

The notation $1,10 = 70$ $1,20 = 80$ $2,10 = 130$ etc. is "sexagesimal" in the sense that 60 units of one kind are written as 1 of the next higher order. This is exactly the same principle we found in Ptolemy's table of chords. The only difference consists in the fact that Old-Babylonian texts have not yet developed a special sign for "zero". This appears, however, in both mathematical and astronomical cuneiform texts of the Seleucid period, as we shall see in later examples. Thus we have reached complete identity of the principle of numerical notation for astronomical tables of the Hellenistic period, whether written in cuneiform or in Greek alphabetic numerals. Only in one point is the Greek notation less consistent than the Babylonian method. In the latter all numbers were written strictly sexagesimally, regardless of whether they were integers or fractions. In Greek astronomy, however, only the fractions were written sexagesimally, whereas for integer degrees or hours the ordinary alphabetic notation remained in use also for numbers from 60 onwards.

Thus Ptolemy would write 130 17 20 where a cuneiform tablet would have 2 10 17 20. In other words, the Greeks already introduced the inconsistency which is still visible in modern astronomy, where one also would write 130°17′20″. The other inconstistency of the modern astronomical notation, namely, to continue beyond the seconds with decimal fractions, is a recent invention. It is interesting to see that it took about 2000 years of migration of astronomical knowledge from Mesopotamia via Greeks, Hindus, and Arabs to arrive at a truly absurd numerical system.

13. The example of our present system of numeration for degrees, hours, measures and ordinary numbers should suffice totally to discredit the popular idea that a number system was "invented" at a certain moment. Yet innumerable "reasons" have been advanced why the Babylonians used the basis 60 for their number system. I shall not make any attempt to discuss here the history of the sexagesimal system in any detail, but a few points must be mentioned because they are of importance for the historical approach to the development of number systems as a whole.

First of all, there exists a common misconception as to the generality of the use of the sexagesimal system. The very same tablet which contains hundreds of sexagesimal numbers, column beside column, to compute the dates of the new moons for a given year, might end with a "colophon" containing the name of the owner of the tablet, the name of the scribe, and the date of writing of the text, the year being expressed in the form 2 *me* 25 "2 hundred 25" where the main text would express the very same date sexagesimally as 3,45. In other words, it is only in strictly mathematical or astronomical contexts that the sexagesimal system is consistently applied. In all other matters (dates, measures of weight, areas, etc.), use was made of mixed systems which have their exact parallel in the chaos of 60-division, 24-division, 12-division, 10-division, 2-division which characterizes the units of our own civilization. The question of the origin of the sexagesimal system is therefore inextricably related to the much more complex problem of the history of many concurrent numerical notations and their innumerable local and chronological variations.

But it is not enough to realize that the 60-division is only one of several contemporary norms between higher and lower units. The essential point lies in the use of the place value notation, regardless of the value of the ratio between consecutive units. No historical theory of the origin of the sexagesimal system is acceptable if it does not account also for this extraordinary feature, namely, the use of the same small number of symbols for different values, depending on the arrangement. A variety of "bases" is well known from number words and number writing all over the world. The place value notation, however, is the most striking feature of the Babylonian system.

A problem of this kind cannot be solved by speculation, but only by a systematic analysis of the written documents. Fortunately there is an abundance of source material available. The early association of Assyriology with Biblical problems and the Hellenistic and Roman concept of "Chaldaeans" as equivalent to astrologers or magicians is today still reflected in the widespread idea that the majority of Babylonian documents are concerned with religion, magic or number mysticism. In fact, however, the overwhelming majority of cuneiform texts concern economic items. Tens of thousands of such documents were unearthed and, although only a small fraction has been made available in modern publications, they suffice to obtain a fair sampling of the use of numbers through all periods of Mesopotamian history. Especially for the earliest period of writing the economic records are almost the only class of existing documents and the number signs are among those signs which one can read with certainty even for periods where the interpretation of the other signs is very problematic.

In the earliest phase of writing the signs are still recognizable as pictures which were scratched in the soft clay with the sharpened edge of a stylus, probably made of reed. The number signs, however, were impressed with the round end of the stylus. A slanted position of the stylus produces a roughly ellipse-shaped impression; a vertical position results in a circular sign. The former represents ordinary units, the latter tens. Thus we can read a 40 in the second section from the top in the second column from the left in Pl. 4b. In the first column we have a 2 in the 7th section from the top. The subsequent sections are damaged but one can still recognize the traces of 7, 8, and 9 in their respective compartments preceding a 10 in the next to the last section at the left lower corner.

Beside these basic elements, many modifications of number symbols were in use for different classes of objects, such as capacity measures, weights, areas, etc. Among these a clear decimal system has been recognized with signs for 1, 10, and 100. The numbers 1 and 10 we have already described. The 100 was written as a circular impression which looks like 10, but is made much bigger. Thus 100 is simply "big 10". Another system proceeds sexagesimally, at least partially. Distinct units are 1 and 10 as before. A big 1 represents 60. Two big units written in opposing directions are combined into one sign to form 120. A 10-sign added in the middle gives 1200. A very big 10 sign stands for 3600. Variations of these systems, both decimal and more or less sexagesimal, can be established at different localities. The main facts, however, are common to all of them, namely, the existence of a decimal substratum and the use of bigger symbols to represent higher units. This latter fact is obviously the root for the development of the place value notation.When the script slowly became simplified and standardized, the distinction between bigger and smaller signs of the same type disappeared. Whereas originally one big unit, meaning 60, and one 10 symbol were written to denote 60 + 10, later a simple "1" followed by a 10 was read 70, in contrast to a 10 followed by 1 meaning 11.

Combined with this, another process was taking place. In economic texts units of weight, measuring silver, were of primary importance. These units seem to have been arranged from early times in a ratio 60 to 1 for the main units "mana" (the Greek μνᾶ "mina") and shekel. Though the details of this process cannot be described accurately, it is not surprising to see this same ratio applied to other units and then to numbers in general. In other words, any sixtieth could have been called a shekel because of the familiar meaning of this concept in all financial transactions. Thus the "sexagesimal" order eventually became the main numerical system and with it the place value writing derived from the use of bigger and smaller signs. The decimal substratum, however, always remained visible for all numbers up to 60.

Similarly, other systems of units were never completely extin-
guished. Only the purely mathematical texts, which we find well
represented about 1500 years after the beginning of writing, have
fully utilized the great advantage of a consistent sexagesimal
place value notation. Again 1000 years later, this method became
the essential tool in the development of a mathematical astronomy,
whence it spread to the Greeks and then to the Hindus, who
contributed the final step, namely, the use of the place value
notation also for the smaller decimal units. It is this system that
we use today.

14. The Babylonian place value notation shows in its earlier
development two disadvantages which are due to the lack of a
symbol for zero. The first difficulty consists in the possibility
of misreading a number 1 20 as 1,20 = 80 when actually
1,0,20 = 3620 was meant. Occasionally this ambiguity is overcome by separating the two numbers very clearly if a whole
sexagesimal place is missing. But this method is by no means
strictly applied and we have many cases where numbers are
spaced widely apart without any significance. In the latest period,
however, when astronomical texts were computed, a special
symbol for "zero" was used. This symbol also occurs earlier as
a separation mark between sentences, and I therefore transcribe
it by a "period." Thus we find in Seleucid astronomical texts
many instances of numbers like 1,.,20 or even 1,.,.,20 which
apply exactly the same principle as, e. g., our 201 or 2001.

But even in the final phase of Babylonian writing we do not
find any examples of zero signs at the end of numbers. Though
there are many instances of cases like .,20 there is no safe example of a writing like 20,. known to me. In other words, in all
periods the context alone decides the absolute value of a sexagesimally written number. In Old-Babylonian mathematical texts we
find several cases where a final result was written by means of
individual symbols for the fractions, e. g., 1,30 might be called
"1 and $\frac{1}{2}$" which shows that we should transcribe $1;30 = 1\frac{1}{2}$ and
not 1,30 = 90.

The ambiguity with respect to fractions and integers is of no
importance for the practice of computation. Exactly as we multiply
two numbers regardless of the position of the decimal point, one
can also operate with the Babylonian numbers and determine

the absolute value at the end if necessary. For the numerical process itself it is indeed a great advantage that one does not need to worry about special values for fractions and integers. It is precisely this feature which gave the Babylonian system its tremendous advantage over all other number systems in antiquity. Though this will become more obvious in the subsequent discussion of (and comparison between) Babylonian and Egyptian mathematics, one example may be given now to illustrate this point.

A multiplication by 12 would be performed by an Egyptian scribe in two steps. First he would multiply the other factor by 10 (simply by replacing each individual symbol by the next higher one) and then he would double the other factor. Finally he would add the two results. Thus for the multiplication of 12 by 12 he would arrange his figures as follows:

	1	12
/	10	120
/	2	24
	total	144

giving him 144 as the result of the addition of the two items marked by a stroke. Let us now assume that the other factor was a fraction, say the "unit fraction" $\frac{1}{5}$, or, as we should write in imitation of the Egyptian notation, $\overline{5}$. The scribe would again proceed in two steps, namely, multiplication by 10 and by 2. The first gives the result 2. The second, however, would need a table of duplications of unit fractions where the double of $\overline{5}$ appears to be listed as $\overline{3}\ \overline{15}$ (indeed $\frac{2}{5} = \frac{1}{3} + \frac{1}{15}$). Thus the computation would be

	1		$\overline{5}$
/	10		2
/	2	$\overline{3}$	$\overline{15}$
	total	2 $\overline{3}$	$\overline{15}$

A contemporary Old-Babylonian scribe would solve the same problems by using a multiplication table for 12 exactly of the same type as we have described above, p. 16, for 10. In line 12 he would directly find the result 2,24. Of course, so far we have only established the fact of a better organized procedure in

Mesopotamia but nothing intrinsically inherent in the Babylonian notation. This is different for the second problem, however. The Babylonian scribe would know (or take this information from a table of reciprocals) that $\frac{1}{5}$ corresponds to "12" ($0;12 = \frac{12}{60}$ in our notation when we use a zero symbol). Hence $\frac{12}{5}$ leads again to finding the value of 12 times 12 or again to 2,24 (we would write 2;24). In other words the Babylonian process completely avoids special rules for computing with fractions, whether unit fractions or not, and requires only that one remember correctly the place value of each contributing number, exactly as we must do in placing the final decimal point. The historical consequences of this simplification can scarcely be overestimated.

14 a. The advantages of the Babylonian place value system over the Egyptian additive computation with unit fractions are so obvious that the sexagesimal system was adopted for all astronomical computations not only by the Greek astronomers but also by their followers in India and by the Islamic and European astronomers. Nevertheless the sexagesimal notation is rarely applied with the strictness with which it appears in the cuneiform texts of the Seleucid period in Mesopotamia. Ptolemy, for example, uses the sexagesimal place value system exclusively for fractions but not for integers. Thus he will write 365 as $\tau\ \xi\ \varepsilon$ (300, 60, 5) but not as $\varsigma\ \varepsilon$ (6,5). The same procedure was followed by the Islamic astronomers and is the reason for our present astronomical custom to write integers decimally and then use sexagesimal minutes and seconds.

Extreme consistency in the use of the sexagesimal place value system is found in the Latin version of the "Alfonsine Tables" (about 1280). Here we find a date like 1477 Sept. 20 $6;1,36^{\text{h}}$ expressed as 2,29,49,32;15,4,0 days. Indeed 1476 Julian years (of $365\frac{1}{4}$ days each) contain $24,36 \cdot 6,5;15$ days $= 2,29,45,9$ days. To this are added the 4,23 days until Sept. 20 and the fraction $0;15,4^{\text{d}} = 6;1,36^{\text{h}}$. This gives the above total of days, counted from A.D. 1 Jan. 0.

Also Copernicus often used consistently written sexagesimal numbers, particularly in his tables of mean motions. For example, for the moon he gives the following mean motions in consecutive Egyptian years (of 365 days each)

1   2,9;37,22,36°
2   4,19;14,45,12
3   0,28;52,7,49   etc.

where we (and Ptolemy) would write for the integers 129, 259, and 28 respectively.

The perfection to which Islamic scholars developed numerical methods has only recently become clear, especially through the work of P. Luckey on al-Kāshī, the astronomer royal of Ulūgh Beg in Samarqand. Al-Kāshī died in 1429; one of his last works is a treatise on the circumference of the circle in which he determines (correctly) 2π as 6;16,59,28,1,34,51,46,15,50. And since he had invented, a few years earlier, the decimal analog of the sexagesimal fractions, he also converts the above number into decimal fractions: 6.2831853071795865.

BIBLIOGRAPHY TO CHAPTER I

The best existing book on numbers and number systems is Karl Menninger, Zahlwort und Ziffer. Aus der Kulturgeschichte unserer Zahlsprache, unserer Zahlschrift und des Rechenbretts. Breslau, Hirt, 1934. The sections about the Babylonian number system are not always reliable. In general, however, this work is far superior to the majority of books on the history of numbers and of elementary computing.

Kurt Sethe, Von Zahlen und Zahlworten bei den alten Aegyptern und was für andere Völker und Sprachen daraus zu lernen ist. Schriften d. wiss. Ges. Strassburg 25 (1916). This work is fundamental for the understanding of the role of fractions within number systems in general.

For calendar computation in general see F. K. Ginzel, Handbuch der mathematischen und technischen Chronologie, 3 vols., Leipzig, Hinrichs, 1906–1914. The lunar calendar of Western Europe during the Middle Ages is discussed in vol. III.

An excellent introduction to Babylonian civilization is given in Edward Chiera, They Wrote on Clay, Univ. of Chicago Press; several editions since 1938.

NOTES AND REFERENCES TO CHAPTER I

ad 1. The "Book of the Hours" of the Duke of Berry was originally published by Paul Durrieu, Les très riches heures de Jean de France, Duc de Berry, Paris 1904. The twelve calendar miniatures are reproduced in color in Verve

No. 7 (1940), unfortunately excluding the zodiacal figure ("melothesia") which followed the calendar. It was discussed in detail by Harry Bober, The Zodiacal Miniature of the Très Riches Heures of the Duke of Berry — Its Sources and Meaning, Journal of the Warburg and Courtauld Institutes 11 (1948) p. 1–34. *ad 2.* D. E. Smith and L. C. Karpinski, The Hindu-Arabic Numerals, Boston, 1911. Julius Ruska, Zur ältesten arabischen Algebra und Rechenkunst, Sitzungsber. d. Heidelberger Akad. d. Wiss., philos.-hist. Kl. 1917, 2. G. F. Hill, The Development of Arabic Numerals in Europe exhibited in 64 Tables. Oxford, Clarendon Press, 1915.

It is a mistake to assume that the Islamic mathematicians and astronomers consistently used the "Hindu-Arabic" numerals. By and large the Hindu-Arabic numerals are restricted to mathematical context, whereas astronomical tables use the alphabetic numerals. In Egypt the Greek or Coptic alphabetic numerals remained in use for centuries after the Arabic conquest.

A nice detail about the transmission of the numerical notation of the Hindus to the Islamic world is accidentally preserved in the autobiography of Ibn Sīnā, the "Avicenna" of the Middle Ages. He was born in 980 near Bukhara, which was then under the Iranian Dynasty of the Samanids. When he was about ten years old, missionaries of an Islamic sect, called Ismaelites, came to Bukhara from Egypt. Through the teaching of these missionaries Ibn Sīnā learned about the Hindu method of computing. Without this explicit bit of information nobody would have dreamed that Indian influence reached southern Russia via Egypt! (Cf. Ibn Sīnā's autobiography, translated in Arthur J. Arberry, Avincenna on Theology, London 1951.)

ad 3. For the development of the Greek number system and its relation to Phoenicia cf. Wilhelm Larfeld, Griechische Epigraphik, 3rd ed., München, Beck, 1914 [Handbuch der klassischen Altertumswissenschaft vol. 1]. See esp. p. 290 ff. Furthermore M. N. Tod, The alphabetic numeral system in Attica. The Annual of the British School at Athens, 45 (1950) p. 126–139.

The acrophonic numerals are often called "Herodianic" because a grammarian Herodianus (second century A. D.) discussed these numbers. The name seems to have been introduced by Woisin in his thesis, De graecorum notis numeralibus, Lipsia 1886.

Recent discussion, textual evidence and bibliography in Marcus Niebuhr Tod, The Greek acrophonic numerals, The Annual of the British School at Athens No. 37, Sessions 1936–37, p. 236–258 (London 1940). For examples cf. B. D. Meritt–H. T. Wade-Gery-M. F. McGregor, The Athenian Tribute Lists (Cambridge, Harvard Univ. Press, 1939) vol. I passim; e. g. the photo p. 74 Fig. 98 and corresponding copy on Pl. XXI.

ad 6. W. E. van Wijk, Le nombre d'or. Étude de chronologie technique suivie du texte de la Massa Compoti d'Alexandre de Villedieu. La Haye, Nijhoff, 1936. This work contains a valuable introduction to the medieval cyclic calendars in Europe. Cf. also Nils Lithberg, Computus, Stockholm 1953 (Swedish) = Nordiska Museets Handlingar 29; very complete bibliography.

ad 7. Jean-Gabriel Lemoine, Les anciens procédés de calcul sur les doigts en orient et en occident. Revue des études islamiques 6 (1932) p. 1–58 [with extensive critical bibliography].

Egyptian numbering of fingers: Kurt Sethe, Ein altägyptischer Fingerzählreim. Zeitschr. für Aegyptische Sprache 54 (1918) p. 16–39. Battiscombe Gunn, "Finger-Numbering" in the Pyramid Texts. ibid. 57 (1922) p. 71 f.

For the relations between alphabetic numerals, number mysticism, astrology, etc., see Franz Dornseiff, Das Alphabet in Mystik und Magie, Stoicheia 7, 2nd ed., Leipzig, Teubner, 1925.

ad 8. Examples for Athenian coins of the New Style: Bulletin de Corre-spondance Hellénique 58 (1934) Pl. I. These coins show the Athenian owl standing on the Panathenaic amphora. The numerals for the months are often inscribed on the amphora and are therefore called "amphora letters". For Greek coins in general see, e. g., Barkley V. Head, Historia Numorum, A Manual of Greek Numismatics, Oxford, Clarendon 1911.

For the Ptolemaic coins with double letter numerals cf. Reginald Stuart Poole, Catalogue of Greek Coins, The Ptolemies, Kings of Egypt, p. 44 and Pl. VIII, 5. These numbers seem to represent the years of an era in honor of Queen Arsinoë II (270 B.C.): cf. Head, Hist. Num. p. 850.

ad 9. The Greek symbols for 6, 90, and 900 are usually called stigma, qoppa, and sampi respectively. The first is originally ϝ (or similar) and therefore also called "digamma", that is double-gamma. Later, it assumed forms which were similar to the ligature of c and τ in Byzantine manuscripts and it was therefore called "stigma" (since the 7th or 8th cent. A.D.). Its original name is Waw. Qoppa is the Q of the Phoenician alphabet. The sampi is originally written with only one middle stroke (cf. the form in Pl. 5). The name "sampi" has been in use since the 17th century A.D.; the Phoenician original is an S-sound called Ṣade.

The division of the circumference of the circle into 360 parts originated in Babylonian astronomy of the last centuries B.C. The sexagesimal number system as such is many centuries older and has nothing to do with astronomical concepts.

ad 10. For an example of an inscription with large number symbols in the alphabetic notation see Inscriptiones Graecae vol. 12,1 (Insularum Maris Aegaei) Berlin 1895, No. 913. This inscription, found in Keskinto (Rhodes) and dating from the second century B.C., lists the basic numbers of a theory of planetary motion; the author is unknown. For a discussion cf. P. Tannery, Mémoires scientifiques vol. 2 p. 487 ff.

In the ordinary alphabetic notation the numbers 1000, 2000 etc. are written by means of α, β, etc. which precede the symbols of lower order. Often accents are added in order to avoid confusion with 1, 2, etc. Several cases, both from inscriptions and papyri, are known, where the symbol for 900, the "sampi", with α, β, ... as superscript was used for 1000, 2000, etc. (cf. Larfeld, quoted above p. 24 in the note to Section 3, p. 294).

In papyri of the early Ptolemaic period one finds, in addition to sampi, also $\varphi v = 500 \; (+) \; 400$ for 900, and besides ω also $\varphi \tau = 500 \; (+) \; 300$ for 800. Cf. Mahaffy, Flinders Petrie Papyri vol. 3 p. 98 etc. An example from a school-book of the third century B.C. is shown on Pl. 5 from P. Cairo. Inv. 65445 (published by O. Guéraud et P. Jouguet, Publications de la Société Royale Égyptienne de Papyrologie, Textes et Documents, Vol. 2, Cairo 1938). The column on the left and the middle column constitute a table of squares of which the following part is clearly readable:

6	6	36	100	100	1 · 10000
7	7	49	200	200	4 · 10000
8	8	64	300	300	9 · 10000
9	9	81	400	400	16 · 10000
10	10	100	500	500	25 · 10000
20	20	400	600	600	36 · 10000
30	30	900	700	700	49 · 10000
40	40	1600	800	800	64 · 10000

Note in Pl. 5 the signs for 6 and 900. The sign for 1000 (in 1600) is an α with an attached loop. The multiples of 10000 are written as a μ (first letter of the Greek word for 10000) with the factor written over it. The last column gives a list of the fractions of the drachma; preserved are the symbols for the following unit fractions: $\bar{8}$, $\overline{12}$, $\overline{24}$, $\bar{3}$, $\bar{6}$, $\bar{\bar{3}}$, $\overline{48}$. Note that the unit fractions are written with the ordinary number signs plus an accent. The only exception is β' which does not mean $\frac{1}{2}$ but $\frac{2}{3}$ denoted here by $\bar{\bar{3}}$. Its corresponding drachma symbol is a combination of the symbols for $\bar{2}$ and $\bar{6}$. Indeed, $\bar{\bar{3}} = \bar{2} + \bar{6}$.

Ordinarily the arrangement of the alphabetic numerals is strictly from higher to lower numbers. In datings, however, one finds also the inverted order: cf. for examples from Mesopotamia Yale Classical Studies 3 p. 30 ff. (clay bullae from Uruk); Excavations in Dura-Europus, Preliminary Report IX, 1 p. 169 ff.; Klio 9 p. 353. For Macedonian inscriptions (between 131 B.C. and 322 A.D.) cf., e. g., Tod, The Macedonian Era; The Annual of the British School at Athens, No. 23 (1918–1919) p. 206–217 and No. 24 (1919–1921) p. 54–67.

That the Arabic form for the zero symbol (a little circle with a bar over it and related forms) is simply taken from Greek astronomical manuscripts was recognized by F. Woepcke in 1863 (Journal Asiatique, Sér. 6 vol. 1 p. 466 ff.). A table showing different forms in Arabic manuscripts as well as in Greek papyri is given by Rida A. K. Irani, Arabic numeral forms, Centaurus 4 (1955) p. 1–12. In a Byzantine manuscript, written about 1300 A.D. a sign like Ч is used for zero beside $\bar{o}$ (Vat. Graec. 1058 fol. 261 ff.), apparently under Islamic influence.

ad 13. The most comprehensive collection of the evidence on early number signs is found in the first edition of Anton Deimel, Šumerische Grammatik der archaistischen Texte, Roma, Pontificium Institutum Biblicum, 1924 (Chapter IV). More recent evidence, especially concerning the decimal system, is given in A. Falkenstein, Archaische Texte aus Uruk, Leipzig 1936, (sign list at the end). For the picture of a stylus see, e. g., S. Langdon, Excavations at Kish, vol. 1, Paris 1924, Pl. XXIX and Falkenstein, l. c., p. 6.

The texts from Uruk also revealed the existence of a system of fractions strictly proceeding on the principle of repeated halving. A very important feature of cuneiform numerical notation is the existence of special signs for $\frac{1}{2}$, $\frac{1}{3}$, $\frac{2}{3}$, and $\frac{5}{6}$ which are in very common use also in later periods, even occasionally in mathematical texts. These "natural fractions" undoubtedly play an important role in the arrangement of metrological units. Obviously one will group higher units in such a form that they admit directly the forming of these most common parts. This leads naturally to a grouping in 12 or 30 or 60. All these ratios do occur in one or another of the parallel systems of units in Mesopotamian metrology.

A strictly decimal notation occurs occasionally in mathematical texts. The following is an example from an Old-Babylonian text (published N e u g e b a u e r-S a c h s, Math. Cuneiform Texts, p. 18). The number 1,12 which occurs in the text is transcribed in the heading as "4 thousand 3 hundred and 20" which is indeed the equivalent of 1,12,0. This example shows at the same time the lack of an absolute determination of the place value in this period of number writing. We may interpret 1,12 as $1,12,0 = 4320$ or as $1,12 = 72$ or as $1;12 = 1\frac{1}{5}$ etc. Only the context permits the determination of the absolute value of a number written sexagesimally.

The lack of a notation which determines the absolute value of a number made it possible to misinterpret simple tables of multiplication or reciprocals. When Hilprecht, in 1906, published a volume of "mathematical, metrological and chronological tablets from the Temple Library of Nippur" he was convinced that these texts showed a relation to Plato's number mysticism. In book VIII of the "Republic" Plato gives some cabbalistic rules as to how guardians of his dictatorially ruled community should arrange for proper marriages. By some wild artifices, Plato's cabbala was brought into relationship with the numbers found on the tablets. Thus 1,10 (i. e. 70 or $1\frac{1}{6}$ etc.) was interpreted to mean 195,955,200,000,000 and in this fashion whole tablets were transcribed and "explained".

As to the origin of the sexagesimal place value notation, it may be noted that it is quite common that fractions of monetary units came to mean fractions in general. As an example can be quoted the Roman *as*, which is $\frac{1}{12}$ of the *uncia* (ounce). In the measurement of time, however, *as* is $\frac{1}{12}$ of one hour (Jahreshefte d. oesterreichischen archaeol. Inst. in Wien 37, 1948, p. 111).

Mixed writings are also quite common. An example from an astronomical procedure text (ACT No. 811a, obv. 27) is 1 *me* 1,30 *me* for "190 days". Here 1 *me* means "1 hundred" (*me* being an abbreviated writing of the Babylonian word for 100), while 1,30 is the sexagesimal writing for 90, and the final *me* means "day", probably an abbreviation.

ad 14. No definite answer can be given to the question when the zero sign was introduced in Babylonian mathematics. We feel sure that it did not exist, say, before 1500: and we find it in full use from 300 B.C. on. A table of squares, found at Kish, tentatively dated by the excavator, S. L a n g d o n, to the period of Darius (500 B.C.), contains four cases of a "zero" written exactly like 30. It is omitted in one case. Cf. Neugebauer, MKT I p. 73 and II pl. 34. For the possibility of an earlier date (about 700) see MCT p. 34 note 95.

One might expect that the Babylonian notation should often lead to errors, e. g., by mistaking 10,2 for 12 and vice versa. Numbers of this type are, however, ordinarily written with very careful spacing such that ⟨𝗍 is hardly ever to be taken as 𝗍𝗍 . The descriptions of the Babylonian number systems in the current textbooks are generally quite misleading on this very point. Nevertheless, there do exist cases where the proper combination of tens and units becomes very doubtful. We even have examples of large numbers, written in two or more lines, where, e. g., the 50 of a 56 was written at the end of one line and the 6 at the beginning of the next. For such "split writings" cf. N e u g e b a u e r-S a c h s, Math. Cun. Texts p. 13 note 69 and N e u g e b a u e r, ACT vol. 2, index.

In practical computations, the Babylonian scribes occasionally committed the same type of mistakes which arise when we are careless with the decimal point. As an example may be quoted an astronomical text which concerns the risings and settings of Mercury during the years 146 to 122 B.C. (BM 34585, Neugebauer ACT No. 302 obv. IV, 30). The scribe had trouble with interpolations. The table at his disposal contained the entries

15	42
45	36

The problem consists in interpolating the value of the right-hand column for the value 31;20 of the left-hand argument. Obviously the answer should be $36 + 13;40 \cdot 0;12 = 36 + 2;44 = 38;44$. The astronomical problem required the addition of this result to another number 1;20. Hence the final result should be 40;4. In the text, however, we find 37;22,44. Obviously the scribe incorrectly determined the place value of $13;40 \cdot 0;12$ and wrote 0;2,44 instead of 2;44. This gave him as the result of the interpolation 36;2,44 and therefore as the final answer $1;20 + 36;2,44 = 37;22,44$.

It must be said, however, that the number of errors in the texts is comparatively very small. I have had the experience that I committed many more errors in checking the ancient computations than there were in the original documents. Often errors in a text are very helpful because they constitute one of the main tools for establishing the details of a numerical procedure followed by the ancient computer.

ad 14. Paul Luckey, Der Lehrbrief über den Kreisumfang (ar-risāla al-muhīṭīya) von Ğamšīd b. Mas'ūd al-Kāšī. Abh. d. Deutschen Akad. d. Wiss. zu Berlin. Kl. f. Math. u. allgem. Naturwiss. Jahrg. 1950 Nr. 6, Akad. Verl., Berlin 1953. Furthermore: Paul Luckey, Die Rechenkunst bei Ğamšīd. b. Mas'ūd al-Kāšī mit Rückblicken auf die ältere Geschichte des Rechnens. Abh. f. d. Kunde d. Morgenlandes 31,1 (1951).

CHAPTER II

Babylonian Mathematics.

15. The following chapter does not attempt to give a history of
Babylonian mathematics or even a complete summary of its
contents. All that it is possible to do here is to mention certain
features which might be considered characteristic of our present
knowledge.

I have remarked previously that the texts on which our study
is based belong to two sharply limited and widely separated
periods. The great majority of mathematical texts are "Old-
Babylonian"; that is to say, they are contemporary with the
Hammurapi dynasty, thus roughly belonging to the period from
1800 to 1600 B.C. The second, and much smaller, group is
"Seleucid", i. e. datable to the last three centuries B.C. These
dates are arrived at on quite reliable palaeographic and linguistic
grounds. The more than one thousand intervening years influenced
the forms of signs and the language to such a degree that one is
safe in assigning a text to either one of the two periods.

So far as the contents are concerned, little change can be
observed from one group to the other. The only essential progress
which was made consists in the use of the "zero" sign in the
Seleucid texts (cf. p. 20). It is further noticeable that numerical
tables, expecially tables of reciprocals, were computed to a much
larger extent than known from the earlier period, though no new
principle is involved which would not have been fully available
to the Old-Babylonian scribes. It seems plausible that the
expansion of numerical procedures is related to the development
of a mathematical astronomy in this latest phase of Mesopotamian
science.

For the Old-Babylonian texts no prehistory can be given.
We know absolutely nothing about an earlier, presumably

Sumerian, development. All that will be described in the subsequent sections is fully developed in the earliest texts known.
It is customary to postulate a long development which is supposedly necessary to reach a high level of mathematical insight.
I do not know on what experience this judgment is based.
All historically well known periods of great mathematical discoveries have reached their climax after one or two centuries of
rapid progress following upon, and followed by, many centuries
of relative stagnation. It seems to me equally possible that Babylonian mathematics was brought to its high level in similarly
rapid growth, based, of course, on the preceding development
of the sexagesimal place value system whose rudimentary forms
are already attested in countless economic texts from the earliest
phases of written documents.

16. The mathematical texts can be classified into two major
groups: "table texts" and "problem texts". A typical representative
of the first class is the multiplication table discussed above p. 16.
The second class comprises a great variety of texts which are all
more or less directly concerned with the formulation or solution
of algebraic or geometrical problems. At present the number
of problem texts known to us amounts to about one hundred
tablets, as compared with more than twice as many table texts.
The total amount of Babylonian tablets which have reached
museums might be estimated to be at least 500,000 tablets and
this is certainly only a small fraction of the texts which are still
buried in the ruins of Mesopotamian cities. Our task can therefore properly be compared with restoring the history of mathematics from a few torn pages which have accidentally survived
the destruction of a great library.

17. The table texts allow us to reconstruct a small, however
insignificant, bit of historical information. The archives from
the city of Nippur, now dispersed over at least three museums,
Philadelphia, Jena, and Istanbul, have given us a large percentage
of table texts, many of which are clearly "school texts", i. e.,
exercises written by apprentice scribes. This is evident, e. g.,
from the repetition in a different hand of the same multiplication
table on obverse and reverse of the same tablet. Often we also
find vocabularies written on one side of a tablet which shows
mathematical tables on the other side. These vocabularies are

the backbone of the scribal instruction, necessary for the mastery of the intricacies of cuneiform writing in Akkadian as well as in Sumerian. Finally, many of our mathematical tables are combined with tables of weights and measures which were needed in daily economic life. There can be little doubt that the tables for multiplication and division were developed simultaneously with the economic texts. Thus we find explicitly confirmed what could have been concluded indirectly from our general knowledge of early Mesopotamian civilization.

18. Though a single multiplication table is rather trivial in content, the study of a larger number of these texts soon revealed unexpected facts. Obviously a complete system of sexagesimal multiplication tables would consist of 58 tables, each containing all products from 1 to 59 with each of the numbers from 2 to 59. Thanks to the place value notation such a system of tables would suffice to carry out all possible multiplications exactly as it suffices to know our multiplication table for all decimal products. At first this expectation seemed nicely confirmed except for the unimportant modification that each single tablet gave all products from 1 to 20 and then only the products for 30, 40, and 50. This is obviously nothing more than a space saving device because all 59 products can be obtained from such a tablet by at most one addition of two of its numbers. But a more disturbing fact soon became evident. On the one hand the list of preserved tables showed not only grave gaps but, more disconcertingly, there turned up tables which seemed to extend the expected scheme to an unreasonable size. Multiplication tables for 1,20 1,30 1,40 3,20 3,45 etc. seemed to compel us to assume the existence not of 59 single tables but of 3600 tables. The absurdity of this hypothesis became evident when tables for the multiples of 44,26,40 repeatedly appeared; obviously nobody would operate a library of $60^3 = 216000$ tablets as an aid for multiplication. And it was against all laws of probability that we should have several copies of multiplication tables for 44,26,40 but none for 11, 13, 14, 17, 19 etc.

The solution of this puzzle came precisely from the number 44,26,40 which also appears in another type of tables, namely, tables of reciprocals. Ignoring variations in small details, these tables of reciprocals are lists of numbers as follows

2	30	16	3,45	45	1,20
3	20	18	3,20	48	1,15
4	15	20	3	50	1,12
5	12	24	2,30	54	1,6,40
6	10	25	2,24	1	1
8	7,30	27	2,13,20	1,4	56,15
9	6,40	30	2	1,12	50
10	6	32	1,52,30	1,15	48
12	5	36	1,40	1,20	45
15	4	40	1,30	1,21	44,26,40

The last pair contains the number 44,26,40 and also all the
other two-place numbers mentioned above occur as numbers
of the second column. On the other hand, with one single exception to be mentioned presently, the gaps in our expected list of
multiplication tables correspond exactly to the missing numbers
in our above table of reciprocals. Thus our stock of multiplication tables is not a collection of tables for all products $a \cdot b$, for
a and b from 1 to 59, but tables for the products $a \cdot \bar{b}$ where $\bar{b}$
is a number from the right-hand side of our last list. The character
of these numbers $\bar{b}$ is conspicuous enough; they are the reciprocals
of the numbers b of the left column, written as sexagesimal fractions:

$$\tfrac{1}{2} = 0;30$$
$$\tfrac{1}{3} = 0;20$$
$$\tfrac{1}{4} = 0;15$$
$$\text{etc.}$$
$$\tfrac{1}{1,21} = 0;0,44,26,40.$$

We can express the same fact more simply and historically more
correctly in the following form. The above "table of reciprocals"
is a list of numbers, b and $\bar{b}$, such that the products $b \cdot \bar{b}$ are
1 or any other power of 60. It is indeed irrelevant whether we
write

$$2 \cdot 30 \ \ = 1,0$$

or

$$2 \cdot 0;30 = 1$$

or

$$0;2 \cdot 30 \ \ = 1$$

or

$$0;2 \cdot 0;30 = 0;1 \quad \text{etc.}$$

Experience with the mathematical problem texts demonstrates in innumerable examples that the Babylonian mathematicians made full use of this flexibility of their system.

Thus we have seen that the tables of multiplication combined with the tables of reciprocals form a complete system, designed to compute all products $a \cdot \bar{b}$ or, as we now can write, all sexagesimal divisions $\dfrac{a}{b}$ within the range of the above-given table of reciprocals. This table is not only limited but it shows gaps. There is no reciprocal for 7, for 11, for 13 or 14, etc. The reason is obvious. If we divide 7 into 1 we obtain the recurrent sexagesimal fraction 8,34,17,8,34,17, ...; similarly for $\frac{1}{11}$ the group 5,27,16,21,49 appears in infinite repetition. We have tables which laconically remark "7 does not divide", "11 does not divide", etc. This holds true for all numbers which contain prime numbers not contained in 60, i. e. prime numbers different from 2, 3, and 5. We shall call these numbers "irregular" numbers in contrast to the remaining "regular" numbers whose reciprocals can be expressed by a sexagesimal fraction of a finite number of places.

We have mentioned one exception to our rule that all multiplication tables must concern numbers $\bar{b}$ or, as we shall call them now, regular numbers. This is the case of the first irregular number, namely, 7, for which several multiplication tables are preserved. The purpose of this addition is clearly the completion of all tables $a \cdot b$ at least for the first decade, in which 7 would be the only gap because all the remaining numbers from 1 to 10 are regular. Thus we see that our original assumption was correct for the modest range from 1 to 10. Instead, however, of expanding this table up to 60, one chooses a much more useful sequence of numbers, namely, those which are needed not only for multiplication but also for division. The mere multiplications could always be completed by one simple addition from two different tables. This system of tables alone, as it existed in 1800 B.C., would put the Babylonians ahead of all numerical computers in antiquity. Between 350 and 400 A.D., Theon Alexandrinus wrote pages of explanations in his commentaries to Ptolemy's sexagesimal computations in the Almagest. A scribe of the administration of an estate of a Babylonian temple 2000 years before Theon would have rightly wondered about so many words for such a simple technique.

The limitations of the "standard" table of reciprocals which we reproduced above (p. 32) did not mean that one could not transgress them at will. We have texts from the same period teaching how to proceed in cases not contained in the standard table. We also have tables of reciprocals for a complete sequence of consecutive numbers, regular and irregular alike. The reciprocals of the irregular numbers appear abbreviated to three or four places only. But the real expansion came in the Seleucid period with tables of reciprocals of regular numbers up to 7 places for b and resulting reciprocals up to 17 places for $\overline{b}$. A table of this extent, containing the regular numbers up to about $17 \cdot 10^{12}$, can be readily used also for determining approximately the reciprocals of irregular numbers by interpolation. Indeed, in working with astronomical texts I have often used this table exactly for this purpose and I do not doubt that I was only repeating a process familiar to the Seleucid astronomers.

19. Returning to the Old-Babylonian period we find many more witnesses of the numerical skill of the scribes of this period. We find tables of squares and square roots, of cubes and cube roots, of the sums of squares and cubes needed for the numerical solution of special types of cubic equations, of exponential functions, which were used for the computation of compound interest, etc.

Very recently A. Sachs found a tablet which he recognized as having to do with the problem of evaluating the approximation of reciprocals of irregular numbers by a finite expression in sexagesimal fractions. The text deals with the reciprocals of 7, 11, 13, 14, and 17, in the last two cases in the form that $b \cdot \overline{b} = 10$ instead of $b \cdot \overline{b} = 1$ as usual. We here mention only the two first lines, which seem to state that

$$8,34,16,59 < \overline{7}$$

but

$$8,34,18 > \overline{7}.$$

Indeed, the correct expansion of $\overline{7}$ would be 8,34,17 periodically repeated. It is needless to underline the importance of a problem which is the first step toward a mathematical analysis of infinite arithmetical processes and of the concept of "number" in general. And it is equally needless to say that the new fragment raises many more questions than it solves. But it leaves no doubt that we must recognize an interest in problems of approximations for as early a period as Old-Babylonian times.

This is confirmed by a small tablet, now in the Yale Babylonian Collection (cf. Pl. 6a). On it is drawn a square with its two diagonals. The side shows the number 30, the diagonal the numbers 1,24,51,10 and 42,25,35. The meaning of these numbers becomes clear if we multiply 1,24,51,10 by 30, an operation which can be easily performed by dividing 1,24,51,10 by 2 because 2 and 30 are reciprocals of one another. The result is 42,25,35. Thus we have obtained from $a = 30$ the diagonal $d = 42;25,35$ by using

$$\sqrt{2} = 1;24,51,10.$$

The accuracy of this approximation can be checked by squaring 1;24,51,10. One finds

$$1;59,59,59,38,1,40$$

corresponding to an error of less than $22/60^4$. Expressed as a decimal fraction we have here the approximation 1.414213.. instead of 1.414214... This is indeed a remarkably good approximation. It was still used by Ptolemy in computing his table of chords almost two thousand years later.

Another Old-Babylonian approximation of $\sqrt{2}$ is known to be 1;25. It is also contained in the approximation of $\sqrt{2}$ which we find in the Hindu Śulva-Sūtras whose present form might be dated to the 3rd or 4th century B.C. There we find

$$\sqrt{2} = 1 + \frac{1}{3} + \frac{1}{3 \cdot 4} - \frac{1}{3 \cdot 4 \cdot 34}$$

whose sexagesimal equivalent is

$$1;25 - 0;0,8,49,22,\ldots = 1;24,51,10,37,\ldots.$$

The possibility seems to me not excluded that both the main term and the subtractive correction are ultimately based on the two Babylonian approximations.

20. The above example of the determination of the diagonal of
the square from its side is sufficient proof that the "Pythagorean"
theorem was known more than a thousand years before Pythagoras.
This is confirmed by many other examples of the use of this
theorem in problem texts of the same age, as well as from the
Seleucid period. In other words it was known during the whole
duration of Babylonian mathematics that the sum of the squares
of the lengths of the sides of a right triangle equals the square of
the length of the hypotenuse. This geometrical fact having once
been discovered, it is quite natural to assume that all triples of
numbers l, b, and d which satisfy the relation $l^2 + b^2 = d^2$ can
be used as sides of a right triangle. It is furthermore a normal
step to ask the question: When do numbers l, b, d satisfy the
above relation? Consequently it is not too surprising that we find
the Babylonian mathematicians investigating the number-theoret-
ical problem of producing "Pythagorean numbers". It has often
been suggested that the Pythagorean theorem originated from
the discovery that 3, 4, and 5 satisfy the Pythagorean relation.
I see no motive which would lead to the idea of forming triangles
with these sides and to investigate whether they are right triangles
or not. It is only on the basis of our education in the Greek
approach to mathematics that we immediately think of the
possibility of a geometric representation of arithmetical or algebraic relations.

To say that the discovery of the geometrical theorem led naturally to the corresponding arithmetical problem is very different
from expecting that the latter problem was actually solved. It is
therefore of great historical interest that we actually have a text
which clearly shows that a far reaching insight into this problem
was obtained in Old-Babylonian times. The text in question
belongs to the Plimpton Collection of Columbia University in
New York.

As is evident from the break at the left-hand side, this tablet
was originally larger; and the existence of modern glue on the
break shows that the other part was lost after the tablet was
excavated. Four columns are preserved, to be counted as usual
from left to right. Each column has a heading. The last heading
is "its name" which means only "current number", as is evident
from the fact that the column of numbers beneath it counts simply

the number of lines from "1st" to "15th". This last column is therefore of no mathematical interest. Columns II and III are headed by words which might be translated as "solving number of the width" and "solving number of the diagonal" respectively. "Solving number" is a rather unsatisfactory rendering for a term which is used in connection with square roots and similar operations and has no exact equivalent in our modern terminology. We shall replace these two headings simply by "*b*" and "*d*" respectively. The word "diagonal" occurs also in the heading of the first column but the exact meaning of the remaining words escapes us.

The numbers in columns I, II and III are transcribed in the following list. The numbers in [ ] are restored. The initial numbers "[1]" in lines 4 ff. are half preserved, as is clearly seen from the photograph (Pl. 7 a). A "1" is completely preserved in line 14. In the transcription I have inserted zeros where they are required; they are not indicated in the text itself.

I	II (= *b*)	III (= *d*)	IV
[1,59,0,]15	1,59	2,49	1
[1,56,56,]58,14,50,6,15	56,7	3,12,1	2
[1,55,7,]41,15,33,45	1,16,41	1,50,49	3
[1,]5[3,1]0,29,32,52,16	3,31,49	5,9,1	4
[1,]48,54,1,40	1,5	1,37	5
[1,]47,6,41,40	5,19	8,1	6
[1,]43,11,56,28,26,40	38,11	59,1	7
[1,]41,33,59,3,45	13,19	20,49	8
[1,]38,33,36,36	9,1	12,49	9
1,35,10,2,28,27,24,26,40	1,22,41	2,16,1	10
1,33,45	45	1,15	11
1,29,21,54,2,15	27,59	48,49	12
[1,]27,0,3,45	7,12,1	4,49	13
1,25,48,51,35,6,40	29,31	53,49	14
[1,]23,13,46,40	56	53	15

This text contains a few errors. In II,9 we find 9,1 instead of 8,1 which is a mere scribal error. In II,13 the text has 7,12,1 instead of 2,41. Here the scribe wrote the square of 2,41, which is 7,12,1 instead of 2,41 itself. In III,15 we find 53 instead of 1,46 which is twice 53. Finally there remains an unexplained error in III,2 where 3,12,1 should be replaced by 1,20,25.

The relations which hold between these numbers are the following ones. The numbers b and d in the second and third columns are Pythagorean numbers; this means that they are integer solutions of

$$d^2 = b^2 + l^2$$

As b and d are known from our list, we can compute l and find

Line	l	Line	l	Line	l
1	2,0	6	6,0	11	1,0
2	57,36	7	45,0	12	40,0
3	1,20,0	8	16,0	13	4,0
4	3,45,0	9	10,0	14	45,0
5	1,12	10	1,48,0	15	1,30

If we then form the values of $\dfrac{d^2}{l^2}$ we obtain the numbers of column I. Thus our text is a list of the values of $\dfrac{d^2}{l^2}$, b, and d, for Pythagorean numbers. It is plausible to assume that the values of l were contained in the missing part. That they have been explicitly computed is obvious.

If we take the ratio $\dfrac{b}{l}$ for the first line we find $\dfrac{1,59}{2,0} = 0;59,30$ that is, almost 1. Hence the first right triangle is very close to half a square. Similarly one finds that the last right triangle has angles close to 30° and 60°. The monotonic decrease of the numbers in column I suggests furthermore that the shape of the triangles varies rather regularly between these two limits. If one investigates this general fact more closely, one finds that the values of $\dfrac{d^2}{l^2}$ in column I decrease almost linearly and that this holds still more accurately for the ratios $\dfrac{d}{l}$ themselves (Fig. 3).

This observation suggests that the ancient mathematician who composed this text was interested not only in determining triples of Pythagorean numbers but also in their ratios $\dfrac{d}{l}$. Let us investigate the mathematical character of this problem. We know that all Pythagorean triples are obtainable in the form

$$l = 2pq \qquad b = p^2 - q^2 \qquad d = p^2 + q^2$$

where p and q are arbitrary integers subject only to the condition that they are relatively prime and not simultaneously odd and $p > q$. Consequently we obtain for the ratio $\dfrac{d}{l}$ the expression

$$\frac{d}{l} = \tfrac{1}{2}(p \cdot \bar{q} + \bar{p} \cdot q)$$

where $\bar{p}$ and $\bar{q}$ are the reciprocals of p and q. This shows that $\dfrac{d}{l}$ are expressible as finite sexagesimal fractions, as is the case in our text, if and only if both p and q are regular numbers.

This fact can be easily checked in our list of numbers by computing the values of p and q which correspond to the l, b, and d of our text. Then one finds a very remarkable fact. The numbers p and q are not only regular numbers, as expected, but they are regular numbers contained in the "standard table"

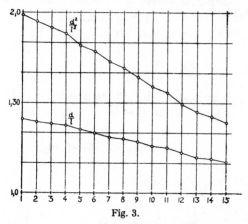

Fig. 3.

of reciprocals (p. 32) so well known to us from many tables of the same period. The only apparent exception is $p = 2,5$ but this number is again well known as the canonical example for the computation of reciprocals beyond the standard table. This

seems to me a strong indication that the fundamental formula
for the construction of triples of Pythagorean numbers was
known. Whatever the case may be, the text in question remains
one of the most remarkable documents of Old-Babylonian
mathematics. We shall presently (p. 42) return to the question
how a formula for Pythagorean numbers could have been found.

21. Pythagorean numbers were certainly not the only case of
problems concerning relations between numbers. The tables for
squares and cubes point clearly in the same direction. We also
have examples which deal with the sum of consecutive squares
or with arithmetic progressions. It would be rather surprising if
the accidentally preserved texts should also show us the exact
limits of knowledge which were reached in Babylonian mathematics. There is no indication, however, that the important
concept of prime number was recognized.

All these problems were probably never sharply separated
from methods which we today call "algebraic". In the center of
this group lies the solution of quadratic equations for two unknowns. As a typical example might be quoted a problem from a
Seleucid text. This problem requires the finding of a number such
that a given number is obtained if its reciprocal is added to it.

Using modern notation we call the unknown number x, its
reciprocal $\bar{x}$, and the given number b. Thus we have to determine
x from

$$x\,\bar{x} = 1 \qquad x + \bar{x} = b.$$

In the text b has the value 2;0,0,33,20. The details of the solution
are described step by step in the text as follows. Form

$$\left(\frac{b}{2}\right)^2 = 1;0,0,33,20,4,37,46,40.$$

Subtract 1 and find the square root

$$\sqrt{\left(\frac{b}{2}\right)^2 - 1} = \sqrt{0;0,0,33,20,4,37,46,40} = 0;0,44,43,20.$$

The correctness of this result is checked by squaring. Then add
to and subtract from $\dfrac{b}{2}$ the result. This answers the problem:

$$x = \frac{b}{2} + \sqrt{} = 1;0,0,16,40 + 0;0,44,43,20 = 1;0,45$$

$$\overline{x} = \frac{b}{2} - \sqrt{} = 1;0,0,16,40 - 0;0,44,43,20 = 0;59,15,33,20.$$

Indeed, x and $\overline{x}$ are reciprocal numbers and their sum equals the given number b.

This problem is typical in many respects. It shows, first of all, the correct application of the "quadratic formula" for the solution of quadratic equations. It demonstrates again the unrestricted use of large sexagesimal numbers. Finally, it concerns the main type of quadratic problems of which we have hundreds of examples preserved, a type which I call "normal form": two numbers should be found if (a) their product and (b) their sum or difference is given. It is obviously the purpose of countless examples to teach the transformation of more complicated quadratic problems to this "normal form"

$$x \cdot y = a$$
$$x \pm y = b$$

from which the solution then follows as

$$x = \frac{b}{2} + \sqrt{\left(\frac{b}{2}\right)^2 \mp a}$$

$$y = \pm \frac{b}{2} \mp \sqrt{\left(\frac{b}{2}\right)^2 \mp a}$$

simply by transforming the two original equations into two linear equations

$$x \pm y = b$$
$$x \mp y = \sqrt{b^2 \mp 4a}.$$

In other words, reducing a quadratic equation to its "normal form" means finally reducing it to the simplest system of linear equations.

The same idea can be used for finding three numbers, a, b, c, which satisfy the Pythagorean relation. Assume that one again started from a pair of linear equations

$$a = x + y$$
$$b = x - y$$

realizing that
$$a^2 = b^2 + c^2 \quad \text{if} \quad c^2 = 4xy.$$

Assuming that x and y are integers, then a and b will be integers; but $c = 2\sqrt{xy}$ will be an integer only if $\sqrt{xy}$ is an integer. This condition is satisfied if we assume that x and y are squares of integers
$$x = p^2 \qquad y = q^2$$

and thus we obtain the final result that a, b, and c form a Pythagorean triple if p and q are arbitrary integers $(p > q)$ and if we make
$$a = p^2 + q^2 \qquad b = p^2 - q^2 \qquad c = 2pq.$$

This is indeed the formula which we needed for our explanation of the text dealing with Pythagorean numbers.

22. It is impossible to describe in the framework of these lectures the details of the Babylonian theory of quadratic equations. It is not really necessary anyhow, since the whole material is easily available in the editions quoted in the bibliography to this chapter. A few features of this Babylonian algebra, however, deserve special emphasis because they are essential for the evaluation of this whole system of early mathematics.

First of all, it is easy to show that geometrical concepts play a very secondary part in Babylonian algebra, however extensively a geometrical terminology may be used. It suffices to quote the existence of examples in which areas and lengths are added, or areas multiplied, thus excluding any geometrical interpretation in the Euclidean fashion which seems so natural to us. Indeed, still more drastic examples can be quoted for the disregard of reality. We have many examples concerning wages to be paid for labor according to a given quota per man and day. Again, problems are set up involving sums, differences, products of these numbers and one does not hesitate to combine in this way the number of men and the number of days. It is a lucky accident if the unknown number of workmen, found by solving a quadratic equation, is an integer. Obviously the algebraic relation is the only point of interest, exactly as it is irrelevant for our algebra what the letters may signify.

Another important observation concerns the form in which all these algebraic problems are presented. The texts fall into two

$$x = \frac{b}{2} + \sqrt{\ \ } = 1;0,0,16,40 + 0;0,44,43,20 = 1;0,45$$

$$\bar{x} = \frac{b}{2} - \sqrt{\ \ } = 1;0,0,16,40 - 0;0,44,43,20 = 0;59,15,33,20.$$

major classes. One class formulates the problem and then proceeds to the solution, step by step, using the special numbers given at the beginning. The text often terminates with the words "such is the procedure". The second class contains collections of problems only, sometimes more than 200 on a single tablet of the size of a small printed page. These collections of problems are usually carefully arranged, beginning with very simple cases e. g., quadratic equations in the normal form, and expanding step by step to more complicated relations, but all eventually reducible to the normal form. One standard form of such collections consists in keeping the condition $xy = 10,0$ fixed but varying the second equation to more and more elaborate polynomials, ending up, e. g., with expressions like

$$(3x+2y)^2 + \tfrac{9}{13}\left\{4\left[\tfrac{1}{7}((x+y)-(\tfrac{1}{2}+1)(x-y))\right]^2 + (x+y)^2\right\}$$
$$= 4,45,0.$$

Investigating such series, one finds that they all have the same pair $x = 30$ $y = 20$ as solutions. This indicates that it was of no concern to the teacher that the result must have been known to the pupil. What he obviously had to learn was the method of transforming such horrible expressions into simpler ones and to arrive finally at the correct solutions. We have several tablets of the first class which solve one such example after another from corresponding collections of the second class.

From actually computed examples it becomes obvious that it was the general procedure, not the numerical result, which was considered important. If accidentally a factor has the value 1 the multiplication by 1 will be explicitly performed, obviously because this step is necessary in the general case. Similarly we find regularly a general explanation of the procedure. Where we would write $x + y$ the text would say "5 and 3, the sum of length and width". Indeed it is often possible to transform these examples directly into our symbolism simply by replacing the ideograms which were used for "length", "width", "add", "multiply" by our letters and symbols. The accompanying numbers are hardly more than a convenient guide to illustrate the underlying general process. Thus it is substantially incorrect if one denies the use of a "general formula" to Babylonian

algebra. The sequences of closely related problems and the
general rules running parallel with the numerical solution form
de facto an instrument closely approaching a purely algebraic
operation. Of course, the fact remains that the step to a consci-
ously algebraic notation was never made.

23. The extension of this "Babylonian algebra" is truly remarkable. Though the quadratic equations form obviously the most
significant nucleus a great number of related problems were also
considered. Linear problems for several unknowns are common
in many forms, e. g., for "inheritance" problems where the shares
of several sons should be determined from linear conditions
which hold between these shares. Similar problems arise from
divisions of fields or from general conditions in the framework
of the above mentioned collections of algebraic examples.

On the other hand we know from these same collections
series of examples which are equivalent to special types of equations of fourth and sixth order. Usually these problems are
easily reducible to quadratic equations for x^2 or x^3 but we have
also examples which lead to more general relations of 5th and
3rd order. In the latter case the tables for $n^2 + n^3$ seem to be useful
for the actual numerical solution of such problems, but our source
material is too fragmentary to give a consistent description of
the procedure followed in cases which are no longer reducible to
quadratic equations.

There is finally no doubt that problems were also investigated
which transcend, in the modern sense, the algebraic character.
This is not only clear from problems which have to do with
compound interest but also from numerical tables for the consecutive powers of given numbers. On the other hand we have
texts which concern the determination of the exponents of given
numbers. In other words one had actually experimented with
special cases of logarithms without, however, reaching any
general use of this function. In the case of numerical tables the
lack of a general notation appears to be much more detrimental
than in the handling of purely algebraic problems.

24. Compared with the algebraic and numerical component
in Babylonian mathematics the role of "geometry" is rather
insignificant. This is, in itself, not at all surprising. The central
problem in the early development of mathematics lies in the

numerical determination of the solution which satisfies certain conditions. At this level there is no essential difference between the division of a sum of money according to certain rules and the division of a field of given size into, say, parts of equal area. In all cases exterior conditions have to be observed, in one case the conditions of the inheritance, in another case the rules for the determination of an area, or the relations between measures or the customs concerning wages. The mathematical importance of a problem lies in its arithmetical solution; "geometry" is only one among many subjects of practical life to which the arithmetical procedures may be applied.

This general attitude could be easily exemplified by long lists of examples treated in the preserved texts. Most drastically, however, speak special texts which were composed for the use of the scribes who were dealing with mathematical problems and had to know all the numerical parameters which were needed in their computations. Such lists of "coefficients" were first identified by Professor Goetze of Yale University in two texts of the Yale Babylonian Collection. These lists contain in apparently chaotic order numbers and explanatory remarks for their use. One of these lists begins with coefficients needed for "bricks" of which there existed many types of specific dimensions, then coefficients for "walls", for "asphalt", for a "triangle", for a "segment of a circle", for "copper", "silver", "gold", and other metals, for a "cargo boat", for "barley", etc. Then we find coefficients for "bricks", for the "diagonal", for "inheritance", for "cut reed" etc. Many details of these lists are still obscure to us and demonstrate how fragmentary our knowledge of Babylonian mathematics remains in spite of the many hundreds of examples in our texts. But the point which interests us here at the moment becomes very clear, namely, that "geometry" is no special mathematical discipline but is treated on an equal level with any other form of numerical relation between practical objects.

These facts must be clearly kept in mind if we nevertheless speak about geometrical knowledge in Babylonian mathematics, simply because these special facts were eventually destined to play a decisive role in mathematical development. It must also be underlined that we have not the faintest idea about anything amounting to a "proof" concerning relations between geometrical

magnitudes. Several tablets dealing with the division of areas
show figures of trapezoids or triangles but without any attempt at
being metrically correct. The description of geometry as the
science of proving correct theorems from incorrect figures certainly fits Babylonian geometry so far as the figures are concerned
and also with regard to the algebraic relations. But the real
"geometric" part often escapes us. It is, for instance, not at all
certain whether the triangles and trapezoids are right-angle
figures or not. If the texts mention the "length" and "width" of
such a figure it is only from the context that we can determine
the exact meaning of these two terms. If the area of a triangle is
found by computing $\frac{1}{2} a \cdot b$ it is plausible to assume that a and b
are perpendicular dimensions, but there exist similar cases where
only approximate formulae seem equally plausible.

There are nevertheless cases where no reasonable doubt can
arise as to the correct interpretation of geometrical relations.
The concept of similarity is utilized in numerous examples. The
Pythagorean theorem is equally well attested; the same holds
for its application to the determination of the height of a circular
segment. On the other hand only a very crude approximation
for the area of a circle is known so far, corresponding to the use
of 3 for π. Several problems concerning circular segments and
similar figures are not yet fully understood and it seems to me
quite possible that better approximations of π were known and
used in cases where the rough approximation would lead to
obviously wrong results.

As in the case of elementary areas similar relations were known
for volumes. Whole sections of problem texts are concerned with
the digging of canals, with dams and similar works, revealing
to us exact or approximate formulae for the corresponding
volumes. But we have no examples which deal with these objects
from a purely geometrical point of view.

24 a. After completion of the manuscript, new discoveries were
made which must be mentioned here because they contribute
very essentially to our knowledge of the mathematics of the Old-
Babylonian period. In 1936 a group of mathematical tablets
were excavated by French archaeologists at Susa, the capital of
ancient Elam, more than 200 miles east of Babylon. A preliminary
report was published in the Proceedings of the Amsterdam

Academy by E. M. Bruins in 1950 and the following remarks are based on this preliminary publication, though I restrict myself to the most significant results only. The texts themselves still remain unpublished, more than 20 years after their discovery.

The main contribution lies in the direction of geometry. One tablet computes the radius r of a circle which circumscribes an isosceles triangle of sides 50, 50, and 60 (result $r = 31;15$). Another tablet gives the regular hexagon, and from this the approximation $\sqrt{3} \approx 1;45$ can be deduced. The main interest, however, lies in a tablet which gives a new list of coefficients similar to those mentioned above, p. 45. The new list contains, among others, coefficients concerning the equilateral triangle (confirming the above approximation $\sqrt{3} \approx 1;45$), the square ($\sqrt{2} \approx 1;25$), and the regular pentagon, hexagon, heptagon, and the circle. If A_n denotes the area, s_n the side of a regular n-gon, then one can explain the coefficients found in the list as follows:

$$A_5 = 1;40 \cdot s_5^2$$
$$A_6 = 2;37,30 \cdot s_6^2$$
$$A_7 = 3;41 \cdot s_7^2.$$

If we, furthermore, call c_6 the circumference of the regular hexagon, c the periphery of the circle, then the text states

$$c_6 = 0;57,36 \cdot c.$$

Because $c_6 = \dfrac{3}{\pi} c$, the last coefficient implies the approximation

$$\pi \approx 3;7,30 = 3\tfrac{1}{8}$$

thus confirming finally my expectation that the comparison of the circumference of the regular hexagon with the circumscribed circle must have led to a better approximation of π than 3.

The relations for A_5, A_6, and A_7 correspond perfectly to the treatment of the regular polygon in Heron's Metrica XVIII to XX, a work whose close relationship to pre-Greek mathematics has become obvious ever since the decipherment of the Babylonian mathematical texts.

Also in many other respects do the tablets from Susa supplement and confirm what we knew from the contemporary Old-Babylonian sources in Mesopotamia proper. One example deals with the division of a triangle into a similar triangle and a trapezoid such that the product of the partial sides and of the partial areas are given values, the hypotenuse of the smaller triangle being known. This is a new variant of similar problems involving sums of areas and lengths or the product of areas. One of the tablets from Susa implies even a special problem of the 8th degree, whereas until now we had only the sixth degree represented in the Babylonian material. The new problem requires that one find the sides x and y of a rectangle whose diagonal is d, such that $xy = 20,0$ and $x^3 \cdot d = 14,48,53,20$. This is equivalent to a quadratic equation for x^4

$$x^8 + a^2 x^4 = b^2$$

$a = 20,0$ $b = 14,48,53,20$. The text proceeds to give the step-by-step solution of this equation, resulting in $x^4 = 11,51,6,40$ and finally leading to $x = 40$ $y = 30$.

25. However incomplete our present knowledge of Babylonian mathematics may be, so much is established beyond any doubt: we are dealing with a level of mathematical development which can in many aspects be compared with the mathematics, say, of the early Renaissance. Yet one must not overestimate these achievements. In spite of the numerical and algebraic skill and in spite of the abstract interest which is conspicuous in so many examples, the contents of Babylonian mathematics remained profoundly elementary. In the utterly primitive framework of Egyptian mathematics the discovery of the irrationality of $\sqrt{2}$ would be a strange miracle. But all the foundations were laid which could have given this result to a Babylonian mathematician, exactly in the same arithmetical form in which it was obviously discovered so much later by the Greeks. And even if it were only due to our incomplete knowledge of the sources that we assume that the Babylonians did not know that $p^2 = 2 q^2$ had no solution in integer numbers p and q, even then the fact remains that the consequences of this result were not realized. In other words Babylonian mathematics never transgressed the threshold of pre-scientific thought. It is only in the last three centuries of Babylonian history and in the field of mathematical astronomy that the Babylonian mathematicians or astronomers reached parity with their Greek contemporaries.

BIBLIOGRAPHY TO CHAPTER II

The description of Babylonian mathematics given here is far from complete, even if measured by the fragmentary existing material. A more detailed description has been given in the author's "Vorgriechische Mathematik", Berlin, Springer 1934. Any serious study must be based, however, on the texts themselves, in order to get a proper estimate of the sometimes fluent boundaries between established facts and modern interpretation. The majority of the now available texts are published in the following works:

O. N e u g e b a u e r, Mathematische Keilschrift-Texte, 3 vols., Berlin, Springer, 1935/37 = Quellen und Studien zur Geschichte der Mathematik, Abt. A, vol. 2. [Henceforth quoted MKT.]

O. N e u g e b a u e r–A. S a c h s, Mathematical Cuneiform Texts. American Oriental Society, New Haven, 1945 = Am. Oriental Series vol. 29. [Henceforth quoted MCT.]

Both these works give photographs, copies, transcriptions, translations, and commentaries. The main part of the material published in the first-mentioned work was republished in transcription and with translation by F. T h u r e a u - D a n g i n, Textes mathématiques babyloniens, Leiden, Brill, 1938 = Ex Oriente Lux vol. 1.

Since the publication of the first edition of this book several more mathematical texts have come to light, supplementing but not essentially changing the general picture given in the preceding pages. Most of these new texts are again Old-Babylonian problem texts, published by Sayyid Taha Baqir in Sumer vol. 6 (1950) p. 39–54, p. 130–148 and vol. 7 (1951) p. 28–45 (found in Tell Ḥarmal, just south of Baghdad). For a list of coefficients and related subjects cf. A. G o e t z e, Sumer 7 (1951) p. 126–154. Additional texts were discussed by B r u i n s in Sumer 9 and 10 (1953/54) but only in exerpts or in very unreliable transcriptions. A problem text of unknown origin, concerning a circular city (similar to MKT I p. 144) was published by W. F. L e e m a n s in Compte Rendu de la seconde rencontre assyriologique intern., Paris 1951, p. 31–35. Two fragments of problem texts and 16 table texts from the Late-Babylonian archive in Babylon are reproduced in Pinches-Strassmaier-Sachs, Late Babylonian Astronomical and Related Texts, Providence, Brown University Press, 1955. Cf. Also S a c h s, J. Cuneiform Studies 6 (1952) p. 151–156.

NOTES AND REFERENCES TO CHAPTER II

ad 17. There exists a single fragment of a mathematical text written in Sumerian (MKT I p. 234 f.). Because Sumerian was still practiced in the schools of the Old-Babylonian period nothing can be concluded from such a text for the Sumerian origin of Mesopotamian mathematics. The same holds for the exceedingly frequent use of Sumerian words and phrases throughout all periods.

That mathematics was taught in scribal schools can hardly be doubted. At what level such instruction started and to what extent it was the common knowledge of scribes it is impossible to say. There exists a text, probably itself written for use in scribal schools, in which the trying life of a schoolboy in such a "Tablet House" is dramatically described. Cf. S. N. Kramer, Schooldays, a Sumerian Composition Relating to the Education of a Scribe. J. Am. Oriental Soc. 69 (1949) p. 199–215.

ad 18. The structure of the system of tables of multiplication and reciprocals was first described by the present author in a series of papers entitled "Sexagesimalsystem und babylonische Bruchrechnung" I–IV published in Quellen und Studien zur Geschichte der Mathematik, Abt. B., vols. 1 and 2 (1930–1932).

The methods for the computation of reciprocals not contained in the standard table were analyzed by A. Sachs, Babylonian Mathematical Texts I, Journal of Cuneiform Studies 1 (1947) p. 219–240. The transformation of sexagesimal fractions to unit fractions was discussed by the same author in "Notes on Fractional Expressions in Old Babylonian Mathematical Texts", J. of Near Eastern Studies 5 (1946) p. 203–214.

That a scribe was sometimes not quite sure when a number was regular or irregular is shown by a statement found in a text now in the British Museum (MKT I p. 224,12 and p. 184), to the effect that "4,3 does not divide". This is wrong because $4,3 = 3 \cdot 1,21$ and both 3 and 1,21 are regular numbers whose reciprocals can be found in the standard table.

ad 19. Approximations of $\sqrt{2}$ can be found as follows. Obviously $\frac{3}{2} = 1;30$ is a first approximation, though larger than the correct value because $\left(\frac{3}{2}\right)^2 = \frac{9}{4} = 2;15$. Consequently we obtain an approximation which is too small by dividing 2 by $\frac{3}{2}$. The result is $\frac{4}{3} = 1;20$. The mean value of these two opposite approximations is 1;25 which is one of the two attested values. We can repeat this process; $1;25^2 = 2;0,25$ is too large. Thus 2 divided by 1;25 or 1;24,42,21, ... is too small. The mean value of 1;25 and 1;24,42,21 is 1;24,51,10 which is the second approximation found in our texts.

We have no proof that this was the way in which these values were found but there is also no way to disprove this possibility. Cf. also below p. 52.

ad 20. In linie 2 of column III of the Plimpton tablet we find for d the value 3,12,1 instead of 1,20,25. Following a suggestion by R. J. Gillings (The Australian Journal of Science 16, 1953, p. 54–56) one might explain this mistake as the result of two errors: In computing

$$d^2 = p^2 + q^2 = (p + q)^2 - 2pq \quad (p = 1,4 \quad q = 27)$$

the scribe replaced $-2pq$ by $+2pq$ and then wrote down only $2 \cdot 27 \cdot 1,0 = 54,0$ instead of $2 \cdot 27 \cdot 1,4 = 57,36$. Thus he found not $d = 2,18,1 - 57,36 = 1,20,25$ but $2,18,1 + 54,0 = 3,12,1$.

A different suggestion was made by E. M. Bruins, in Sumer 11 (1956) p. 117–121 which contains, however, incorrect and unfair statements as to the readings of the text.

It may finally be remarked that the construction of Pythagorean triples by means of two numbers p and q from

$$l = 2pq \qquad b = p^2 - q^2 \qquad d = p^2 + q^2$$

was well known in Hellenistic times. Diophantus, whose relationship to the oriental tradition is obvious, uses it frequently (e. g. VI, 1). The same holds for the Indian mathematicians, e. g., Mahavira (about 850 A.D.) or Bhascara (about 1150); cf. Mahavira, ed. Nangacarya p. 209 and Lilavati VI, 135 and 145, Colebrooke.

ad 22. Examples for nonhomogeneous terms:

MCT p. 74: addition of areas and volumes.

MKT I p. 243; MKT II p. 63: addition of lengths and areas.

MKT II p. 64: addition of length and volume.

MKT I p. 513: addition of number of days and of men.

Thureau-Dangin, TMB p. 209 f.: addition of sheep and rams.

For a careful classification of quadratic equations see Solomon Gandz, The Origin and Development of the Quadratic Equations in Babylonian, Greek, and Early Arabic Algebra. Osiris 3 (1937) p. 405–557.

ad 23. For cubic equations see MKT I p. 208 ff., and vol. III p. 55. Fourth order: MKT I p. 420; p. 456 and III, p. 62; p. 471 ff.; p. 498. Fifth order: MKT I p. 411. Sixth order: MKT I p. 460.

Tables for a^n: MKT I p. 77 ff. For logarithms cf. MKT I p. 362, MCT p. 35. The tables for a^n were found at Kish, east of Babylon, and at Tell Harmal near Baghdad.

An interesting table of special square roots is contained in the following text (MKT III p. 52):

1	e	1	ib–si$_8$
1,2,1	e	1,1	ib–si$_8$
1,2,3,2,1	e	1,1,1	ib–si$_8$
1,2,3,4,3,2,1	e	1,1,1,1	ib–si$_8$

A following "catch line" points to a succeeding table for the cube roots of 1, 1,3,3,1 etc. The knowledge of the binomial coefficients lies, of course, fully within the reach of Babylonian algebra.

ad 24. It is of interest to remark that not only were similar triangles frequently used in the solution of problems which have geometrical background, but that the arithmetical concept "ratio" had a special term. MKT I p. 460 ff. we have series of examples where the "ratio" x/y is to be computed from quadratic equations. That the ratio of two numbers is treated as an entity is indeed a very important step in the development of algebra.

The area A of a circle is usually determined from its circumference c in the form

$$A = 0;5 \cdot c^2$$

where $0;5 = 1/12$ is an approximation of $\dfrac{1}{4\pi}$. For many examples cf. MCT p. 44 or p. 9.

Problems concerning circular segments are published MCT 56, MKT I p. 188. Cf. also MKT I p. 177, p. 230; MCT 134 ff. All these problems cause trouble — which is a certain indication that we have not yet found the proper key to this part of Babylonian geometry. A very interesting early text concerning ornamental patterns of circles and squares was published by J. C. Gadd (cf. MKT I p. 137 ff.). Unfortunately no solutions are given.

For the inaccuracy of figures cf. MCT p. 46 and p. 54. For the approximate determination of volumes see MKT I p. 165, p. 176; for areas MCT p. 46.

For a clear case of a figure which must be exactly a right triangle in order to make the following similarity relations correct, cf. the tablet published by Sayyid Taha Baqir, Sumer 6 (1950) p. 39–54.

ad 24 a. The paper referred to in the text was preceded by a preliminary note in the August 1950 issue of the French popular journal "Atomes. Tous les aspects scientifiques d'un nouvel age", (p. 270 f.). This article also gives a photograph of a triangle with its circumscribed circle.

The value $\sqrt{3} \approx 1;45$ can be obtained immediately by the process described above p. 50 in the case of $\sqrt{2}$. Starting with the obvious estimate $\sqrt{3} \approx 1;30$, one obtains as the next value $\dfrac{3}{1;30} = 2$ and hence $\frac{1}{2}(2 + 1;30) = 1;45$. The same approximation is found in Heron, Metrica XXV (who uses 1;44 also). Metrica XXI contains $\sqrt{2} \approx 1;25$.

The value $\pi \approx 3\frac{1}{8}$ does not seem to be attested in the preserved literature of antiquity. As its first appearance, Tropfke (Geschichte d. Elementarmathematik IV(³) p. 279) quotes a passage from Dürer in 1525. In Babylonian material this value was hesitatingly mentioned as a possibility in MCT p. 59 note 152k. This conjecture is now fully confirmed.

CHAPTER III

The Sources; Their Decipherment and Evaluation.

26. There are many forces which cooperate in the destruction of source material, none more powerful than continuous peaceful life. The lack of interest in the far remote past will invariably change and eventually destroy what remains from earlier generations.Without violent catastrophies there would hardly be any archeology. If Mesopotamian cities had not been turned into desert hills we would have no chance of finding the hundreds of thousands of documents from which Babylonian history is written.

It would be pointless to describe in any detail the unending sequence of disaster which provided us with the material for our studies. We shall devote our discussion exclusively to the modern attempts at reviving the past and penetrating into the intellectual life of previous generations.

Following the general plan of these lectures I shall not try to describe familiar aspects of historical research at great length. I will emphasize, however, some specific methodological facts which are directly related to our main topic, the investigation of ancient science.

27. Best known of all, of course, is Greek science. The great classics are carefully edited and many of them are available in excellent translations. For the historian of science this is very pleasant indeed, but far from sufficient. Nobody would expect a historian of English literature to remain satisfied with an edition of Shakespeare or Chaucer. Also the history of a science can only be written if more is available than the "classics". The predecessors, the pupils, the related authors are still exceedingly difficult to reach.

Let me mention only one very typical example. Ptolemy's

"Almagest" was edited in Greek by Heiberg in 1898–1903. An excellent critical and annotated translation into German by Manitius was published in 1912–1913. But the planned glossary was never printed, making it exceedingly difficult to check Ptolemy's terminology with other sources. This fact is also responsible for serious gaps even in the best modern Greek dictionaries, not to mention that the history of astronomical terminology in general is practically a terra incognita. Modern editions have played havoc with abbreviations, symbols, drawings, etc. in the original manuscripts. One has to do practically all the work over again if one should try to investigate the development of symbols like the zodiacal signs or the planetary sigla. The best list of mathematical, astronomical and chemical symbols is still the collection made by Du Cange in the appendix to his "Glossarium ..." (1688) which, in turn, is based on a sheet, now in the Bibliothèque Nationale, written around 1480 by Angelo Poliziano, the teacher of Piero di Medici. This is a characteristic example of the true state of affairs in the study of the history of scientific developments.

It is only recently that scholars following A. Rome and A. Delatte, have begun to publish editions where the figures and their lettering are taken as part of the text. With these recent exceptions no edition can be trusted in the least with respect to appearance, lettering or even existence of figures. The question, for instance, how the ancients depicted geometrical relations on a sphere cannot be seriously discussed on the basis of the existing printed texts.

Ptolemy's other works are slowly being published. One volume appeared in 1907, containing, among others, important writings on the theory of sun dials and on stereographic projection which is the basis of the famous astrolabe, one of the most important instruments of medieval astronomy. The "Tetrabiblos", the "Bible of astrology", was published twice during World War II. One edition, by E. Boer, appeared in Germany, Greek text only; the other, by F. E. Robbins, Greek with English translation, in the Loeb Classical Library. Thus once the experiment was unwillingly made of testing the uniqueness of the modern text-critical methods. It is amusing to see that the differences begin with the title and continue in varying degree in the division of chapters and sections. Of course in essence the results are the same, but the details are by no means identical.

An enormous literature has clustered around Ptolemy's "Geography", one of the most influential books of antiquity. Nevertheless, no reliable edition exists. The task is indeed of great difficulty. The constant use of this work has greatly affected its tradition and it is a major enterprise to restore the original version of a text which to a large extent consists of geographical names and numbers uncheckable by internal evidence which is fortunately available in the case of astronomical tables.

To summarize, we may say that even Ptolemy's work is only in part available, disregarding completely lost works, fragments of which may or may not appear on papyri or in some obscure oriental library.

Early Greek astronomy from its beginnings about 400 B.C. to Ptolemy (about 150 A.D.) is almost completely destroyed, except for a few very elementary works which survived for teaching purposes. But the rest was obliterated by Ptolemy's outstanding work, which relegated his predecessors to merely historically interesting figures.

As to Ptolemy's successors we should be in a much better position. Pappus's and Theon's commentaries, written in the 4th century, were widely used and have in part survived. They are now in the process of publication by A. Rome. We are still very badly off so far as the tables are concerned, though at least a preliminary publication by Halma exists, more than 100 years old and bristling with misprints and errors. But almost nothing has been done with Byzantine or European medieval tables. Thus all work on mathematical astronomy of the Middle Ages is most seriously handicapped by the fact that almost no tables are accessible, though hundreds of them can be found listed in library catalogs. The much publicized "progress" in the study of the history of science is difficult to reconcile with the shocking neglect of a great wealth of source material which is of primary importance for our knowledge of Byzantine astronomy. The study of the problem of the interaction between Islamic science and the West is precluded as long as these sources remain unpublished. What we really need is not bibliographies and summaries, but competent publications of Islamic, Greek, and Latin treatises.

28. There is one group of sources which will become of increasing importance when systematically utilized: the astrological writings. During the last 60 years a "catalogue" in 12 volumes of Greek astrological writings has been completed. The text is Greek, the notes are Latin, the indices are restricted to proper names and occasionally to selected terminology. And the content can only be called repelling—hundreds and hundreds of pages of the dryest astrological nonsense. Nevertheless I think that the scholars who have undertaken this publication, foremost of all Franz Cumont, have contributed enormously to the study of ancient civilization, far beyond the narrow limits of the history of astrology. To quote only one work, I mention Cumont's "L'Égypte des astrologues" (1937). Here, stripped of all astrology, the background of the daily life and of the contemporary institutions of Hellenistic Egypt is depicted as it becomes visible from the mishaps and fortunes predicted for the men and women who consulted the astrologer. And another fact of great historical importance becomes increasingly clear from these researches, namely, that the date of origin of this mighty astrological lore must be fixed to the Ptolemaic period in Egypt and thus appears as a truly Hellenistic creation.

We shall come back to this aspect of our problem in the last chapter. At the moment we will only consider the astrological literature as a source of information for the history of astronomy. Indeed, these texts contain innumerable scattered fragments of computations concerning the moon, the planets, positions of stars, their risings and settings. These computations are often almost hopelessly distorted. Many centuries of tradition through handwritten copies have badly affected numbers which were of little interest, if not unintelligible to the scribes. Nevertheless, we obtain from these passages many references to methods which belong to the period between Hipparchus and Ptolemy. One of the most unexpected discoveries was made by W. Gundel. In an Old-French and a related Latin astrological manuscript of the Renaissance he found imbedded the fragments of a star catalogue of the time of Hipparchus. Slowly there emerges from scattered scraps of information a whole system of astronomical methods which are very different from the classical "Ptolemaic" system but which are of primary importance for the study of the origin and transmission of Hellenistic astronomy.

29. The majority of manuscripts on which our knowledge of Greek science is based are Byzantine codices, written between

500 and 1500 years after the lifetime of their authors. This suffices for one to realize the importance of every scrap of papyrus from a scientific or astrological treatise. Here we have originals which were written during the Hellenistic period itself, not yet subject to the selective editing of late centuries. It can be said without any exaggeration that the relatively young field of papyrology has truly revolutionized classical studies—even if by natural inertia this has not always become evident in the standard curricula.

The fascinating story of the recovery of papyrus treasures from the soil of Egypt has often been told and need not be repeated here. A masterly little book by one of the leading scholars in the field, H. Idris Bell, "Egypt from Alexander the Great to the Arab Conquest" (Oxford, 1948), will not only give the reader a summary of papyrology, its history and methods, but is itself a brilliant study of the diffusion and decay of Hellenism, the very problem one facet of which is the subject of the present lectures.

Papyrology is one of the best organized and most pleasantly managed fields of the humanities. An unusual spirit of cooperation has survived two great wars. Great series of publications of texts have appeared regularly and the intrinsic difficulties of the field and its enormous spread into highly specialized disciplines, especially law, agriculture, economics, etc. have created the active cooperation of scholars of neighboring disciplines. Consequently, generally usable editions were produced with translations, commentaries and excellent indices, glossaries and handbooks—indeed a pleasant contrast to, say, the fact that the Arabic version of Euclid is published only with a Latin translation (1897–1932) so as to make life miserable to a mathematician who would perhaps for once want to look into the Oriental tradition of a classic in his field.

In spite of the very active and successful work of papyrologists their number is much too small to cope with the large amount of material which has reached museums and smaller collections. Many hundreds of papyri and papyrus fragments are rapidly disintegrating into dust after having been purchased at high prices from antiquity dealers. From my own very limited experience I could quote several instances where papyri got lost in smaller collections, which have neither facilities nor competent personnel for the proper handling of these documents. Especially well preserved texts are valuable for the antiquities trade and are therefore most exposed to rapid destruction. Again I might mention one typical example. One of the most interesting astronomical papyri eventually reached in part the Carlsberg collection of the University of Copenhagen. H. O. Lange and I published this text, which was written probably at the beginning of our era. It contains a Demotic translation and commentary to a much older hieratic text whose hieroglyphic replica is still preserved in the Cenotaph of King Seti I (about 1300 B.C.). One of its subjects is the description of the travel of the "decans" over the body of the sky goddess, who was depicted on the ceilings of tombs and temples as a representation of the vault of heavens under which we live. Our papyrus was first seen in the posession of an antiquities dealer in Cairo. At the time, the text still contained at the beginning the picture of the sky goddess with all the constellations and their dates of rising and setting. When the text reached Copenhagen the picture was gone. No doubt it was sold to some private collector and is probably lost forever. Thus a vital part for the understanding of the text vanished almost at the same moment its importance was recognized.

Needless to say, such happenings are contradictory to existing antiquities laws. Also needless to say, many texts disappear from excavations or are "found" by natives who have long learned that papyri can be sold profitably instead of burning them at their camp fires or using them as fertilizer. But it remains a rather depressing fact that a large percentage of all these spoils is destined to end unread and unpublished in climates less favorable than the soil of Egypt.

30. But all that we have mentioned so far in source material is child's play compared with ancient Mesopotamia. It is barely a hundred years since cuneiform writing once more became intelligible; and it was only shortly before the beginning of the present century that so fundamental a fact as Sumerian being a language of its own though written with the same characters as its later Semitic successors, Akkadian, became generally recognized. But while decipherment and interpretation progressed in slow steps, texts were found in tremendous numbers from the very beginning. The first collection of reliefs and tablets

arrived in France in 1846, having been excavated three years before in the ruins of Khorsabad, near Mosul, by the French consul Botta. In 1849/50 Layard found in the ruins of Nineveh, then called Kuyunjik, the first palace library; in 1853 followed the discovery of the Library of Ashurbanipal by Hormuzd Rassam. Some 20,000 tablets in the British Museum now bear the inventory letter K (for Kuyunjik) or Rm. (for Rassam). Perhaps about a quarter of these two collections is published today.

This ratio between existing and published texts might seem rather small. Actually it is unusually high and only due to the fact that it is the result of a century of work on one of the most famous discoveries in the Near East. In the meantime many tens of thousands of tablets have found their way into museums, providing source material which would require several centuries for their publication even under the concentrated efforts of all living Assyriologists.

31. At this point it may be useful to insert a few general remarks about excavations and publication of texts because little is known about these problems outside the small circle of the initiated.

A modern excavation is a highly complex enterprise. A staff of architects, draftsmen, photographers, epigraphers, and philologists have to assist the archeologist in his field work. But this is only the first and easier phase of an excavation. The preservation of the ruin, the conservation of the objects found, and, most of all, the publication of the results, remains the final task for which the work in the field is only the preliminary step.

Here a sad story indeed must be told. While the field work has been perfected to a very high standard during the last half century, the second part, the publication, has been neglected to such a degree that many excavations of Mesopotamian sites resulted only in a scientifically executed destruction of what was left still undestroyed after a few thousand years. The reasons for this fact are trivial. The time required for the publication of results is a multiple of the requirements for the field work. The available money is usually spent when a fraction of the original planned excavation has been completed, benefactors are hard to find to pay for many years of work without tangible or spectacular results, and the scholars get interested in special aspects

of the problem involved or go out for new material instead of
performing the tedious work of publishing the thousands of
details which the accidents of exavation have provided. The
final result is not much different from the one obtained by the
treasure-hunting attitude of the earliest excavators.

Many an excavation, if not all, had to be stopped before completion or had to restrict itself from the very beginning to a few
trenches crossing the ruin in the hope of getting a general insight
into the character of the stratification. Then one or another
promising building was investigated in greater detail. What
resulted is a ruin left with deep scars, an easy prey for the natives
to extract all exposed bricks, to tunnel for more without too many
difficulties, and to have access to deeper layers and thus to continue the "excavation" in their own fashion and for their own
benefit. In this way the natives must have found thousands of
tablets which were then sold at high prices by antiquities dealers
to the very same museums which spent the initial money for
the removal of many tons of sand and debris.

Let me illustrate the effect upon the special studies under
discussion here. Until 1951 not for a single astronomical or
mathematical text was its provenance established by excavation.
The only apparent exceptions are a number of multiplication
tables from Nippur or Sippar but nobody knows where these
texts were found in the ruins. Consequently it is, e. g., completely
impossible to find out whether these texts came from a temple,
a palace, a private house, etc. Not even the stratum is known to
give us a more accurate date of the texts. In other words, if those
texts, which were officially "excavated", would have been found
by Arabs, we would be no worse off than we are now. But while
the Arabs in their "clandestine" exploits dig only relatively small
holes, a scientific excavation has destroyed beyond any hope all
traces of the locality where the texts have been found. Thus we
are left with the texts alone and must determine their origin from
internal evidence, which is often very difficult to interpret.

A long story could be told about the "methods" to obtain the
needed information. Texts which for more than 50 years were
lying in the basement of a great museum could be relatively
dated from the newspaper in which they were wrapped. That
gave a plausible date for the "expedition" which found the texts
and hence the place from which they were excavated.

A whole class of texts was identified as follows. A German expedition before 1914 had worked in the city of Uruk, a most important site because it contains structures which reach from the earliest periods down to Seleucid times. There the Germans must have found the debris of an archive of which, however, all the good tablets had been removed by the Arabs. These tablets finally found their way into the collections of Berlin, Paris, and Chicago, forming one of the most important groups of texts for the study of Seleucid astronomy. The Arabs were not interested in small fragments. These were left at the site and were then carefully sifted and photographed by the expedition. By courtesy of the Berlin Museum, I obtained prints of these photographs (cf. Pl. 6 b) showing the fragments neatly arranged on a desk of the expedition. The records about the place where they were found were lost in the meantime. The fragments themselves had also been lost. By means of very extensive computation I succeeded, however, in establishing the relationship of these splinters to bigger pieces from the above-mentioned museums. Thus it became possible to restore whole tablets, the parts of which are now on different sides of the Atlantic. Finally, the small fragments themselves were rediscovered in Istanbul. But the main question of their accurate provenance remains unanswerable.

32. The Mesopotamian soil has preserved tablets for thousands of years. This will not be the case in our climate. Many tablets are encrusted with salts (cf. Pl. 9 a left and photograph which shows incrustations along the crack; the right-hand photograph gives the same tablet after cleaning). A change in moisture produces crystals which break the surface of the tablets, thus rapidly obliterating the writing. I have seen "tablets" which consisted of dust only, carefully kept in showcases. To prevent this, tablets must be slowly baked at high temperature and thereafter soaked to remove salts. But only great museums possess the necessary equipment and experienced staffs, not to mention the fact that these methods of conservation were often kept as museum secrets. Many thousands of tablets have been acquired at high cost by big and small collections only to be destroyed without ever being read or recorded in any way.

The publication of tablets is a difficult task in itself. First of all, one must find the texts which concern the specific field in question. This is by no means trivial. Only minute fractions of

the holdings of collections are catalogued. And several of the
few existing rudimentary catalogues are carefully secluded from
any outside use. I would be surprised if a tenth of all tablets in
museums have ever been identified in any kind of catalogue.
The task of excavating the source material in museums is of
much greater urgency than the accumulation of new uncounted
thousands of texts on top of the never investigated previous
thousands. I have no official records of expenditures for excavations at my disposal, but figures mentioned in the press
show that a preliminary excavation in one season costs about as
much as the salary of an Assyriologist for 12 or 15 years. And
the result of every such dig is frequently many more tablets
than can be handled by one scholar in his lifetime.

There exists no simple method of publication. Photographs
alone are in the majority of cases not sufficient, even if their
cost were not prohibitive. Tablets are often inscribed not only on
both sides but also on the edges. Only multiple photographs
taken with variable directions of light would suffice. Thus cost
and actual need have resulted in the practice of hand copies.
Many different styles of copying were developed by individual
scholars, varying between an almost schematic reproduction of
the signs to a minute reproduction of details. The reader may get
an impression of this situation from Pls. 8 and 9 which show
an ephemeris for Saturn (Seleucid period) and an Old Babylonian mathematical text in hand copies and photographs.

The ideal method of publication would be, of course, direct
copying from the text. In practice this is often excluded by the
scattering of directly related material all over the world. Even
with great experience a text cannot be correctly copied without
an understanding of its contents. Practically no text falls at the
first attempt. Thus repeated collation, joining with other fragments, and comparison with other texts are needed. It requires
years of work before a small group of a few hundred tablets is
adequately published. And no publication is "final". Invariably
a fresh mind will find the solution of a puzzle which escaped
the editor, however obvious it might seem afterwards.

33. The process of decipherment follows no fixed rules. Every
special class of texts requires the slow construction of a technical
dictionary. The results of reading difficult signs and words must
be recorded in card files which allow repeated checking of and

comparison with previously obtained results. Only continued experience leads to a more rapid understanding of a certain type of texts. Though it is impossible to describe this process adequately the special situation which prevails in a specific class of mathematical texts might illustrate the great advantages which one has in reading mathematical texts of this type.

The text from which I shall discuss a few lines is reproduced in Pl. 8 a and Pl. 9 a. At first sight it seems impossible to make any sense out of its numbers, which show no relations which could be explained as the result of consecutive operations. Hence one has to abandon the idea of reading the text as a unit. This is confirmed by the short note at the left lower corner of the reverse. This "colophon" reads: 48 im-šu dub-13-kam-ma. The second part means "13th tablet" and characterizes the text as a part of a series of at least 13 related tablets. The first part must refer to the contents. Sometimes the number of lines is indicated; for this 48 is too small a number. But one can easily check that the total of small boxes of two or three lines of text sepaɪated from one another by horizontal lines amounts to about 40 or 50. This is confirmed by similar texts where the number of im-šu exactly agrees with the number of sections. Thus we know already that each section has to be treated individually. Obviously the shortness of these sections suggests that we are dealing with problems only, not with their solutions worked out in detail. This explains the lack of obvious connection between numbers.

Now we are ready to transcribe a few lines of the text, simply following the general Assyriological rules for the transcription and interpretation of signs.We ignore all difficulties of individual readings but we indicate by [ ] a destroyed section. Then we obtain for the lines from 12 to 17 in the left column of the reverse

. . . .

12	gar-gar uš d[ah]5
13	a-rá 2 e-tab
14	uš dah-ma 1
15	sag dah-ma 35
16	a-rá 2 e-tab
17	dah-ma 50

Several terms can be translated directly; gar and daḫ are words known to indicate addition; a-rá is known from the multiplication tables, corresponding to our "times". The same word occurs, e. g., in each line of the multiplication table of Pl. 4 a. The words uš and sag are known to mean length and width respectively. Because no numbers are directly associated with them, we transcribe them by x and y. The particle -ma represents something like "and thus"; we represent it simply by an equal sign. The phrase e-tab seems to suggest another addition because tab is the counterpart to lal "minus" which we know already (p. 5) from the number sign 20 lal 1 = 19. In order not to complicate our discussion unnecessarily we shall anticipate the result that here the whole phrase a-rá 2 e-tab must mean "multiply by 2" without any reference to addition. This is indeed in line with the original meaning of the sign "tab" which consists of two parallel wedges, thus indicating "duplication". We finally remark that we invert for the sake of convenience the order of "x" and "add" and write $+ x$ instead of $x +$. Then we obtain the following "translation".

$$\begin{array}{rl}
& \cdots\cdots\cdots\cdots \\
12 & ++ x[= \quad]5 \\
\hline
13 & \qquad \cdot 2 \\
14 & \quad + x = 1 \\
\hline
15 & + y = 35 \\
\hline
16 & \quad \cdot 2 \\
17 & + (\quad) = 50 \\
\hline
\end{array}$$

We are now facing a new difficulty. At the beginning we tried to read the text as a unit and found that we had to break it up into single problems. Now we have the single problems but they are obviously too short to make sense. Line 15, e. g., requires that y is added to something and then gives the sum 35. And exactly the same difficulty arises in the other examples. Thus we are compelled to introduce an unknown quantity f, which might depend on x and y, to which all the other quantities are added. Thus we interpret line 15 as

$$f + y = 35.$$

It is plausible to assume for the next example the interpretation

$$2f + (y) = 50$$

because otherwise we would have two new unknowns instead of f and y whereas our interpretation makes the second example the direct continuation of its predecessor. But under this assumption we can determine from these two equations the values of f and y, and find $f = 15$ $y = 20$. We can put this result to an immediate test. Line 12 seems to indicate

$$f + x[= \quad]5$$
$$2f + x = 1$$

Obviously the second relation is impossible for positive numbers because $2f + x$ cannot be 1 if $f + x$ already is at least 5. But here the place value notation comes to our rescue. Instead of "1" we can read $1,0 = 60$. Thus we assume $2f + x = 60$ and using $f = 15$ we obtain $x = 30$. Substituting this into the first equation we obtain $f + x = 45$ in excellent agreement with the traces in line 12. Thus we have reached as a first consequence of our hypothetical interpretation that $x = 30$ $y = 20$ $f = 15$.

Again we are able to test this result. Line 12 is the tail end of a bigger section of 6 lines. Following essentially the same method of decipherment we can translate two sentences as follows

$$\tfrac{1}{11}(2,40 - (x + y))$$

and

$$\tfrac{1}{7}(55 - y).$$

If we here substitute $x = 30$ and $y = 20$ we find that the first expression has the value 10, the second 5. The gar-gar "+" in line 12 connects the two preceding expressions. Thus we find a total $10 + 5$ which is indeed the value 15 of f. Hence we have not only confirmed our results but have also determined f as a function of x and y:

$$f = \tfrac{1}{11}(2,40 - (x + y)) + \tfrac{1}{7}(55 - y).$$

In other words we can now summarize all four problems as follows:

$$f + x = 45$$
$$2f + x = 1,0$$
$$f + y = 35$$
$$2f + y = 50$$

where f stands for the above-mentioned expression. Obviously these equations do not suffice, if one takes them singly, to determine x and y. On the other hand they cannot be used simultaneously because there are too many. Thus one has to look for additional information higher up in the text. Applying exactly the same procedure to preceding sections, one finds a simple scheme. There exist several larger sections which define similar functions g, h, etc., always followed by variations of the above form $g + x$, $2g + x$, etc. At the very beginning, however, we find one more condition which turns out to mean

$$xy = 10,0.$$

This condition is common to the whole text as one relation between x and y. All subsequent sections contain individual linear relations between these unknowns, thus leading to quadratic equations for x and y. We know already that $x = 30$ $y = 20$ are the common solutions. Thus our decipherment is completed.

What I have described here is, of course, a simplified story of what actually happened when texts of this type became known, but the essential steps were precisely the same. In this way it was possible to establish many technical terms. The results can be tested in other classes of texts which contain the details of the working out of given problems. And it is clear that the determination of the meaning of a text is generally the easier the more complicated the mathematical context is because this leaves fewer possibilities for the interpretation of the procedure. A context which contains only a few numbers combined by addition or subtraction is of almost no help in the determination of terminology. The advanced algebraic level of Babylonian mathematics was of the greatest help in its being decoded.

NOTES TO CHAPTER III

Extensive bibliographical references for the edition of Greek mathematical and astronomical works are given in R. C. Archibald, Outline of the History of Mathematics, 6th edition, published as a supplement to the American Mathematical Monthly 56 No. 1 (1949).

ad 27. The history of the zodiacal and planetary symbols is virtually unknown. To my knowledge no study has been made which was based on the evidence from manuscripts or epigraphic representation. No symbols appear in the Greek papyri. It is a widespread but wrong belief that the Egyptian hieroglyph ☉ for the sun was used in ancient astronomical texts to denote the sun. The standard symbol in the Middle Ages is ♂, but never ☉ which is commonly used, however, as the abbreviation for οὐρανός (heaven) or κύκλος[1]) (circle). In the latter usage it is shown in Fig. 3a (line 4) in a manuscript of the 10th or 11th century. The solar symbol is directly above it (line 3).

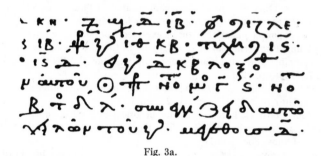

Fig. 3a.

The same holds, to my knowledge, for the papyri (cf. P. Warren 21 from the early third century A.D. and Archiv f. Papyrusforschung 1, 1901, p. 501, a papyrus of the second century A.D.; furthermore Karl Preisendanz, Die griechischen Zauberpapyri, Leipzig 1928–1931, vol. II, index p. 213). Ordinarily, however, the names of sun, moon, and planets are written out in full. Demotic planetary tables of the Roman imperial period contain symbols for the planets and for the zodiacal signs, apparently based on their Egyptian names and with no relation to the mediaeval symbols.

That there is still much left to be done even with the great classics might be illustrated by the fact that Books V to VII of Apollonius's Conic Sections were never edited because they are only preserved in Arabic. The only existing (Latin) translation was made by Edmund Halley in 1710.

One of the many desiderata in the publication of mediaeval tables is the "Alfonsine Tables" which were completed around 1270 under the auspices and active support of Alfonso X (who ruled from 1252 to 1282). Of the Spanish original only the introduction is preserved, but for the Latin versions Haskins (Studies in the History of Medieval Science, p. 17) mentions 75 manuscripts and 13 early editions. Cf. Alfred Wegener, Die astronomischen Werke Alfons X. Bibliotheca Mathematica ser. 3 vol. 6 (1905) p. 129–185.

No accurate estimate can be given about the quality and importance of Byzantine astronomical handbooks but it seems evident from the most superficial use of catalogues that this material must amount to many thousands of folios of texts and tables. It is not known to what extent these treatises continue the

---

[1]) Also in composite words like επι ☉ for ἐπίκυκλος (epicycle).

Ptolemaic-Theonic tradition, are influenced by Islamic works, or have added new ideas of their own.

The wealth of Islamic material is shown by E. S. Kennedy, "A Survey of Islamic Astronomical Tables", Trans. Am. Philos. Soc. N.S. vol. 46,2 (1956) p. 123–177. Of the more than a hundred works listed, only two had been published (al-Battānī by Nallino 1899–1907 and al-Khwārizmī by Bjørnbo, Besthorn, and Suter 1914).

ad 28. The "Catalogus Codicum Astrologorum Graecorum" (Bruxelles, Lamertin, 1898–1953) is not a "catalogue" in the strict sense. The first part of each volume gives a description of the astrological manuscripts. The larger second part contains editions of significant sections of these texts. The volumes are arranged according to countries. Oriental libraries have not been utilized; any amount of source material might be expected from the Near East.

For an important criticism of some results in Cumont's book see L. Robert. Études épigraphiques et philologiques, Bibliothèque de l'école des hautes études 272 (1938) p. 72 ff.

The edition of Ptolemy's astrological writings has been completed by Lammert and Boer in 1952 (Opera III, 2 Teubner).

An edition of Greek horoscopes with historical and astronomical commentaries by the present writer and H. B. Van Hoesen is to appear in the Memoirs of the Am. Philos. Soc. (1957).

Gundel's discoveries were published in "Neue astrologische Texte des Hermes Trismegistos", Abh. d. Bayerischen Akad. d. Wissenschaften, Philos.-histor. Abt., N.F. 12 (1936). The results concerning the old star catalogue are discussed on p. 131–134 and presented in the list of p. 148–153. Gundel determined the date of the catalogue by comparing the given longitudes with the longitudes of the star catalogue in the Almagest and assuming a change of 1° per century, accepting Ptolemy's constant of precession. Though Gundel's main result, namely, that a large number of the positions indicate the time of Hipparchus, cannot be doubted, one can raise objections against details of the discussion. With a single exception (No. 63), all longitudes in Gundel's list are integer degrees (No. 63 gives $29\frac{1}{2}$) and are thus obviously rounded-off values. Ptolemy's catalogue, however, gives longitudes to an accuracy of 10'. Thus one cannot compare these two sets of values without allowance for the rounding-off errors. Two possibilities must be considered; either rounding off to the nearest integer or simple cancellation of fractions. Experience with very large numbers of rounded-off values in Babylonian and Greek astronomy suggests, contrary to our habit, preference for the second method. Consequently I compared the longitudes of all stars of Gundel's list with the longitudes which they had at 130 B.C. according to the catalogue of Peters and Knobel (Ptolemy's Catalogue of Stars, Carnegie Institution of Washington, Publication No. 86, 1915; p. 74 ff.). A comparison is possible for 59 values from Gundel's list and the corresponding values for 130 B.C. disregarding fractions, however close to one. For only one star does the longitude appear to be 1° less than expected for 130 B.C.; and for one other star it is 2° greater than expected. For 39 stars or 66 per cent one obtains exactly the same numbers while 18 stars or 30.5 per cent have longitudes 1° greater than found for 130 B.C. In other words 96.5 per cent of the stars of

Gundel's list have longitudes correct for the time from 130 B.C. to 60 B.C. Thus they were taken either from Hipparchus's catalogue itself or from the catalogue of an astronomer of the next generations.[1]) Gundel's hypothesis, however, of a star catalogue which preceded Hipparchus and which gave the positions in ecliptic coordinates is disproved.

The deeply rooted conviction that the Greeks were inclined towards philosophical speculation only, but neglected observations and experiments made it an easily accepted theory to consider Ptolemy's star catalogue as a trivial modification of Hipparchus's catalogue, simply assuming that Ptolemy added 2;40° to Hipparchus's longitudes, in spite of his explicit statement of independent observations. In the meantime, Boll has shown (Bibliotheca Mathematica ser. 3 vol. 2, 1901, p. 185–195) that Hipparchus's catalogue covered only about 850 stars as compared with more than 1000 of Ptolemy's. Finally, Vogt demonstrated (Astron. Nachr. 224, 1925, cols. 17–54) that for 60 stars of Hipparchus's catalogue, only 5 may have been utilized by Ptolemy whereas the majority undoubtedly show independent observations.

Gundel, who overlooked Vogt's paper, still operated under the assumption of a purely schematic relation between the two catalogues. Of Hipparchus's writings, only his Commentary to Aratus is preserved (edited by Manitius, Leipzig 1894, with a German translation). This work is undoubtedly an early work of Hipparchus, written before the discovery of precession. This follows from the fact that the positions of stars are never given in ecliptic coordinates (longitude and latitude) but in a mixed ecliptic-declination system (cf. Fig. 30 p. 184). It was obviously the discovery of precession that later led Hipparchus to introduce real ecliptic coordinates because longitudes increase proportionally with time whereas latitudes remain unchanged.

ad 29. H. O. Lange–O. Neugebauer, Papyrus Carlsberg I. Ein hieratisch-demotischer kosmologischer Text. Kgl. Danske Vidensk. Selskab, Hist.-filol. Skrifter 1, No. 2 (1940). A new edition with many improvements in detail is in preparation by R. A. Parker and the present author for inclusion in a larger work on Egyptian astronomical texts.

ad 31. Budge, The Rise and Progress of Assyriology, London 1925, writes (p. 136 f.): "As soon as the dealers and officials in Baghdad knew that Rassam was out of the country they began to make excavations on their own account. They employed the workmen who had been employed by Rassam ... The British Museum bought several collections, and as there was keen competition in Paris and America prices began to soar, and in a short time ... tablets ... for which the finders were paid five piastres each in Baghdad, were fetching £ 4 in London". This concerns the period from 1882 to 1887.

The first reliable information about date and location of mathematical

---

[1]) I suspect that for three consecutive stars of Gundel's list (Nos. 45 to 47) the complete values are preserved in "Hermes, De XV stellis" published by Louis Delatte, Textes latins et vieux français relatifs aux Cyranides (Bibl. Fac. Philos. et Lettres Univ. Liège 93, 1942). There one finds for the first three stars the following coordinates ♈ 15;27 ♈ 27;20 ♉ (Text ♈, out of order!) 9;28 corresponding to Gundel's ♈ 15, ♈ 27, ♉ 9. (Delatte p. 246, 249, 250 respectively.) The list "De XV stellis" shares with Gundel's text the preference for the first half of the zodiac; $\frac{4}{5}$ of the stars of Delatte belong to this semicircle as compared with $\frac{2}{3}$ in Gundel.

problem texts of the Old Babylonian period was given by Taha Baqir in Sumer 7 (1951) p. 28 f. According to the field records of the recent excavations in Tell Harmal these texts come from a private house. The joint expedition of the University Museum of the University of Pennsylvania and the Oriental Institute of the University of Chicago to Nippur has now finally established that the "Tablet Hill" which Hilprecht, the original excavator, thought should represent the "Temple Library", actually belongs to "residential quarters of varying date" (D. E. McCown in J. Near Eastern Studies 11, 1952, p. 175).

ad 33. A more extensive description of the method of decipherment of algebraic problem texts is given in my paper "Der Verhältnisbegriff in der babylonischen Mathematik", Analecta Orientalia 12 (1935) p. 235–257.

CHAPTER IV

Egyptian Mathematics and Astronomy.

34. Of all the civilizations of antiquity, the Egyptian seems to me to have been the most pleasant. The excellent protection which desert and sea provide for the Nile valley prevented the excessive development of the spirit of heroism which must often have made life in Greece hell on earth. There is probably no other country in the ancient world where cultivated life could be maintained through so many centuries in peace and security. Of course not even Egypt was spared from severe outside and interior struggles; but, by and large, peace in Mesopotamia or Greece must have been as exceptional a state as war in Egypt.

It is not surprising that the static character of Egyptian culture has often been emphasized. Actually there was as little innate conservatism in Egypt as in any other human society. A serious student of Egyptian language, art, religion, administration, etc. can clearly distinguish continuous change in all aspects of life from the early dynastic periods until the time when Egypt lost its independence and eventually became submerged in the Hellenistic world.

The validity of this statement should not be contested by reference to the fact that mathematics and astronomy played a uniformly insignificant role in all periods of Egyptian history. Otherwise one should deny the development of art and architecture during the Middle Ages on the basis of the invariably low level of the sciences in Western Europe. One must simply realize that mathematics and astronomy had practically no effect on the realities of life in the ancient civilizations. The mathematical requirements for even the most developed economic structures of antiquity can be satisfied with elementary household arithmetic which no mathematician would call mathematics. On the other hand the requirements for the applicability of mathematics to problems of engineering are such that ancient mathematics fell far short of any practical application. Astronomy on the other hand had a much deeper effect on the philosophical attitude of the ancients in so far as it influenced their picture of the world in which we live. But one should not forget that to a large extent the development of ancient astronomy was relegated to the status of an auxiliary tool when the theoretical aspects of astronomical lore were eventually dominated by their astrological interpretation. The only practical applications of theoretical astronomy may be found in the theory of sun dials and of mathematical geography. There is no trace of any use of spherical astronomy for a theory of navigation. It is only since the Renaissance that the practical aspects of mathematical discoveries and the theoretical consequences of astronomical theory have become a vital component in human life.

35. The fact that Egyptian mathematics did not contribute positively to the development of mathematical knowledge does not imply that it is of no interest to the historian. On the contrary, the fact that Egyptian mathematics has preserved a relatively primitive level makes it possible to investigate a stage of development which is no longer available in so simple a form, except in the Egyptian documents.

To some extent Egyptian mathematics has had some, though rather negative, influence on later periods. Its arithmetic was widely based on the use of unit fractions, a practice which probably influenced the Hellenistic and Roman administrative offices and thus spread further into other regions of the Roman empire, though similar methods were probably developed more or less independently in other regions. The handling of unit fractions was certainly taught wherever mathematics was included in a curriculum. The influence of this practice is visible even in works of the stature of the Almagest, where final results are often expressed with unit fractions in spite of the fact that the computations themselves were carried out with sexagesimal fractions. Sometimes the accuracy of the results is sacrificed in favor of a nicer appearance in the form of unit fractions. And this old tradition doubtless contributed much to restricting the sexagesimal place value notation to a purely scientific use.

36. There are two major results which we obtain from the study of Egyptian mathematics. The first consists in the establishment of the fact that the whole procedure of Egyptian mathematics is essentially additive. The second result concerns a deeper insight into the development of computation with fractions. We shall discuss both points separately.

What we mean by the "additivity" of Egyptian mathematics can easily be explained. For ordinary additions and subtractions nothing needs to be said. It simply consists in the proper collection and counting of the marks for units, tens, hundreds, etc., of which Egyptian number signs are composed. But also multiplication and division are reduced to the same process by breaking up any higher multiple into a sum of consecutive duplications. And each duplication is nothing but the addition of a number to itself. Thus a multiplication by 16 is carried out by means of four consecutive duplications, where only the last partial result is utilized. A multiplication by 18 would add the results for 2 and for 16 as shown in the following example

	1	25
/	2	50
	4	100
	8	200
/	16	400
	total	450

In general, multiplication is performed by breaking up one factor into a series of duplications. It certainly never entered the minds of the Egyptians to ask whether this process will always work. Fortunately it does; and it is amusing to see that modern computing machines have again made use of this "dyadic" principle of multiplication. Division is, of course, also reducible to the same method because one merely asks for a factor which is needed for one given number in order to obtain the second given number. The division of 18 by 3 would simply mean to double 3 until the total 18 can be reached

	1	3
/	2	6
/	4	12
	total	18,

and the result is $2 + 4 = 6$. Of course, this process might not always work so simply and fractions must be introduced. To divide 16 by 3 one would begin again with

$$
\begin{array}{ccc}
/ & 1 & 3 \\
  & 2 & 6 \\
/ & 4 & 12
\end{array}
$$

and thus find $1 + 4 = 5$ as slightly below the requested solution. What is still missing is obviously $16 - 15 = 1$, and to this end the Egyptian computer would state

$$
\begin{array}{cc}
\overline{\overline{3}} & 2 \\
/ \quad 3 & 1
\end{array}
$$

which means that $\overline{\overline{3}}$ of 3 is 2, $\overline{3}$ of 3 is 1 and thus he would find 5 $\overline{3}$ as the solution of his problem.

Here we have already entered the second problem, operations with fractions. As we have said in Chapter I, Egyptian fractions are always "unit fractions", with the sole exception of $\overline{\overline{3}}$ which we always include under this name in order to avoid clumsiness of expression[1]). The majority of these numbers are written by means of the ordinary number signs below the hieroglyph ⏜ "r" meaning something like "part". We write therefore $\overline{5}$ for the expression "5th part" $= \frac{1}{5}$. For $\frac{2}{3}$ we write $\overline{\overline{3}}$ whereas the Egyptian form would be "2 parts" meaning "2 parts out of 3", i. e., $\frac{2}{3}$. There exist special signs for $\frac{1}{2}$ and $\frac{1}{4}$ which we could properly represent by writing "half" and "quarter" but for the sake of simplicity we use $\overline{2}$ and $\overline{4}$ as for all other unit fractions.

We shall not go into the details of the Egyptian procedures for handling these fractions. But a few of the main features must be described in order to characterize this peculiar level of arithmetic. If, e. g., $\overline{3}$ and $\overline{15}$ should be added, one would simply leave $\overline{3}\ \overline{15}$ as the result and never replace it by any symbol like $\frac{2}{5}$. Again $\overline{\overline{3}}$ forms an exception in so far as the equivalence of $\overline{2}\ \overline{6}$ and $\overline{\overline{3}}$ is often utilized.

Every multiplication and division which involves fractions leads to the problem of how to double unit fractions. Here we

---

[1]) We disregard here another "complementary fraction" (that is, a fraction of the form $1 - \overline{n}$) which is written with a special sign, namely $\frac{3}{4}$, because it plays no role in the Egyptian arithmetical procedures known to us.

36. There are two major results which we obtain from the study of Egyptian mathematics. The first consists in the establishment of the fact that the whole procedure of Egyptian mathematics is essentially additive. The second result concerns a deeper insight into the development of computation with fractions. We shall discuss both points separately.

What we mean by the "additivity" of Egyptian mathematics can easily be explained. For ordinary additions and subtractions nothing needs to be said. It simply consists in the proper collection and counting of the marks for units, tens, hundreds, etc., of which Egyptian number signs are composed. But also multiplication and division are reduced to the same process by breaking up any higher multiple into a sum of consecutive duplications. And each duplication is nothing but the addition of a number to itself. Thus a multiplication by 16 is carried out by means of four consecutive duplications, where only the last partial result is utilized. A multiplication by 18 would add the results for 2 and for 16 as shown in the following example

$$
\begin{array}{rr}
1 & 25 \\
/\ 2 & 50 \\
4 & 100 \\
8 & 200 \\
/\ 16 & 400 \\
\text{total} & 450
\end{array}
$$

In general, multiplication is performed by breaking up one factor into a series of duplications. It certainly never entered the minds of the Egyptians to ask whether this process will always work. Fortunately it does; and it is amusing to see that modern computing machines have again made use of this "dyadic" principle of multiplication. Division is, of course, also reducible to the same method because one merely asks for a factor which is needed for one given number in order to obtain the second given number. The division of 18 by 3 would simply mean to double 3 until the total 18 can be reached

$$
\begin{array}{rr}
1 & 3 \\
/\ 2 & 6 \\
/\ 4 & 12 \\
\text{total} & 18,
\end{array}
$$

and the result is $2 + 4 = 6$. Of course, this process might not always work so simply and fractions must be introduced. To divide 16 by 3 one would begin again with

	1	3
/	1	3
	2	6
/	4	12

and thus find $1 + 4 = 5$ as slightly below the requested solution. What is still missing is obviously $16 - 15 = 1$, and to this end the Egyptian computer would state

	$\bar{\bar{3}}$	2
/	3	1

which means that $\frac{2}{3}$ of 3 is 2, $\frac{1}{3}$ of 3 is 1 and thus he would find 5 $\bar{3}$ as the solution of his problem.

Here we have already entered the second problem, operations with fractions. As we have said in Chapter I, Egyptian fractions are always "unit fractions", with the sole exception of $\frac{2}{3}$ which we always include under this name in order to avoid clumsiness of expression[1]). The majority of these numbers are written by means of the ordinary number signs below the hieroglyph $\bigcirc$ "r" meaning something like "part". We write therefore $\bar{5}$ for the expression "5th part" $= \frac{1}{5}$. For $\frac{2}{3}$ we write $\bar{\bar{3}}$ whereas the Egyptian form would be "2 parts" meaning "2 parts out of 3", i. e., $\frac{2}{3}$. There exist special signs for $\frac{1}{2}$ and $\frac{1}{4}$ which we could properly represent by writing "half" and "quarter" but for the sake of simplicity we use $\bar{2}$ and $\bar{4}$ as for all other unit fractions.

We shall not go into the details of the Egyptian procedures for handling these fractions. But a few of the main features must be described in order to characterize this peculiar level of arithmetic. If, e. g., $\bar{3}$ and $\overline{15}$ should be added, one would simply leave $\bar{3}$ $\overline{15}$ as the result and never replace it by any symbol like $\frac{2}{5}$. Again $\bar{\bar{3}}$ forms an exception in so far as the equivalence of $\bar{2}$ $\bar{6}$ and $\bar{\bar{3}}$ is often utilized.

Every multiplication and division which involves fractions leads to the problem of how to double unit fractions. Here we

---

[1]) We disregard here another "complementary fraction" (that is, a fraction of the form $1 - \bar{n}$) which is written with a special sign, namely $\frac{3}{4}$, because it plays no role in the Egyptian arithmetical procedures known to us.

find that twice $\bar{2}$, $\bar{4}$, $\bar{6}$, $\bar{8}$, etc. are always directly replaced by 1, $\bar{2}$, $\bar{3}$, $\bar{4}$, respectively. For twice $\bar{3}$ one has the special symbol $\bar{\bar{3}}$. For the doubling of $\bar{5}$, $\bar{7}$, $\bar{9}$, ... however, special rules are followed which are explicitly summarized in one of our main sources, the mathematical Papyrus Rhind. One can represent these rules in the form of a table which gives for every odd integer n the expression for twice $\bar{n}$.

This table has often been reproduced and we may restrict ourselves to a few lines at the beginning:

n	twice $\bar{n}$	
3	$\bar{2}$	$\bar{6}$
5	$\bar{3}$	$\overline{15}$
7	$\bar{4}$	$\overline{28}$
9	$\bar{6}$	$\overline{18}$
	etc.	

The question arises why just these combinations were chosen among the infinitely many possibilities of representing $2/n$ as the sum of unit fractions.

I think the key to the solution of this problem lies in the separation of all unit fractions into two classes, "natural" fractions and "algorithmic" fractions, combined with the previously described technique of consecutive doubling and its counterpart, consistent halving. As "natural" fractions I consider the small group of fractional parts which are singled out by special signs or special expressions from the very beginning, like $\bar{\bar{3}}$, $\bar{3}$, $\bar{2}$ and $\bar{4}$. These parts are individual units which are considered basic concepts on an equal level with the integers. They occur everywhere in daily life, in counting and measuring. The remaining fractions, however, are the unavoidable consequence of numerical operations, of an "algorism", but less deeply rooted in the elementary concept of numerical entities. Nevertheless there are "algorithmic" fractions which easily present themselves, namely, those parts which originate from consistent halving. This process is the simple analogue to consistent duplication upon which all operations with integers are built. Thus we obtain two series of fractions, both directly derived from the "natural" fractions by consecutive halving. One sequence is $\bar{\bar{3}}$, $\bar{3}$, $\bar{6}$, $\overline{12}$, etc., the other $\bar{2}$, $\bar{4}$, $\bar{8}$, $\bar{16}$ etc. The importance of these two series is apparent everywhere in Egyptian arithmetic. A drastic example has already been quoted above on p. 74 where we found that $\bar{3}$ of 3 was found by stating first that $\bar{\bar{3}}$ of 3 is 2 and only as a second step $\bar{3}$ of 3 is 1. This arrangement $\bar{\bar{3}} \to \bar{3}$ is standard even if it seems perfectly absurd to us. It emphasizes the completeness of the first sequence and its origin from the "natural" fraction $\bar{\bar{3}}$.

If one now wishes to express twice a unit fraction, say $\bar{5}$, as a combination of other fractional parts, then it seems natural again to have recourse to these two main sequences of fractions. Thus one tries to represent twice $\bar{5}$ as the sum of a natural fraction of $\bar{5}$ and some other fraction which must be found in one way or another. At this early stage, some trials were doubtless made until the proper solution was found. I think one may reconstruct the essential steps as follows. We operate with the natural fraction $\bar{3}$, after other experiments (e. g., with $\bar{2}$) have failed. Two times $\bar{5}$ may thus be represented as $\bar{3}$ of $\bar{5}$ or $\overline{15}$ plus a remainder which must complete the factor 2 and which is $1\ \bar{\bar{3}}$. The question of finding $1\ \bar{\bar{3}}$ of $\bar{5}$ now arises. This is done in Egyptian mathematics by counting the thirds and writing their number in red ink below the higher units, in our case

$$1 \quad \bar{\bar{3}} \quad \text{(written in black)}$$
$$.\ \ 3 \quad 2 \quad \text{(written in red)}.$$

This means that 1 contains 3 thirds and $\bar{\bar{3}}$ two thirds. Thus the remaining factor contains a total of 5 thirds. This is the amount of which $\bar{5}$ has to be taken. But 5 fifths are one complete unit and this was a third of the original higher unit. Thus we obtain for the second part simply $\bar{3}$ and thus twice $\bar{5}$ is represented as $\bar{3}\ \overline{15}$. This is exactly what we find in the table.

For the modern reader it is more convenient to repeat these clumsy conclusions with modern symbols though we must remember that this form of expression is totally unhistorical. In order to represent $\frac{2}{5}$ in the form of $\dfrac{1}{m} + \dfrac{1}{x}$ we chose $\dfrac{1}{m}$ as a natural fraction of $\frac{1}{5}$, in this case $\frac{1}{3} \cdot \frac{1}{5} = \frac{1}{15}$. For the remaining fraction we have

$$\frac{1}{x} = \left(1 + \frac{2}{3}\right)\frac{1}{5} = \frac{5}{3} \cdot \frac{1}{-} = \frac{1}{-}.$$

Thus we have the representation

$$\frac{2}{5} = \frac{1}{15} + \frac{1}{3}$$

of the table. In general we have

$$\frac{2}{n} = \frac{1}{3} \cdot \frac{1}{n} + \frac{5}{3} \cdot \frac{1}{n}$$

and the second term on the right-hand side will be a unit fraction when and only when n is a multiple of 5. In other words a trial with the natural fraction $\frac{1}{3}$ will work only if n is a multiple of 5. This is indeed confirmed in all cases available in the table of the Papyrus Rhind which covers all expressions for $\frac{2}{n}$ from $n = 3$ to $n = 101$.

We may operate similarly with the natural fraction $\frac{1}{2}$. Then we have

$$\frac{2}{n} = \frac{1}{2} \cdot \frac{1}{n} + \frac{3}{2} \cdot \frac{1}{n}$$

which shows that we obtain a unit fraction on the right-hand side if n is divisible by 3. For $n = 3$ we obtain

$$\frac{2}{3} = \frac{1}{6} + \frac{1}{2}$$

and this is the relation $\overline{3} = \overline{2}\ \overline{6}$ which we quoted at the beginning. All other cases in the table for $\bar{n}$'s which are multiples of 3 show the same decomposition operating with $\dfrac{1}{2\,n}$ as one term.

It is clear that one can proceed in the same manner by operating with $\overline{4}$, $\overline{8}$, etc. or with $\overline{6}$, $\overline{12}$, etc. In this way, more and more cases of the table can be reached and it seems to me there is little doubt that we have found in essence the procedure which has led to these rules for the replacement of $2\,\bar{n}$ by sums of unit fractions.

37. For our present purposes it is not necessary to discuss in detail all steps in the structure of Egyptian fractional arithmetic. I hope, however, to have made clear the two leading principles, the strict additivity and the extensive use of the "natural fractions".

A few historical remarks must be added. The Papyrus Rhind
is not our only document for the study of Egyptian arithmetic.
The other large text, the Moscow papyrus, agrees with rules
known from the Papyrus Rhind. We have, however, an ostracon
from the early part of the New Kingdom where the duplication
of $\overline{7}$ is given as $\overline{6}$ $\overline{14}$ $\overline{21}$ instead of $\overline{4}$ $\overline{28}$ of the standard rule.
Much more material is available from Demotic and Greek papyri
of the Hellenistic period. Here again, deviations from the earlier
rules can be observed, though the main principle remains the
same. In other words we cannot assume that once and forever
a system of fractional tables was computed and then rigidly
maintained. Obviously several equivalent forms were slowly
developed but without ever seriously transgressing the original
methods. This latter fact is of great historical importance. The
handling of fractions always remained a special art in Egyptian
arithmetic. Though experience teaches one very soon to operate
quite rapidly within this framework, one will readily agree that
the methods exclude any extensive astronomical computations
comparable to the enormous numerical work which one finds
incorporated in Greek and late Babylonian astronomy. No wonder
that Egyptian astronomy played no role whatsoever in the development of this field.

38. It would be quite out of proportion to describe Egyptian
geometry here at length. It suffices to say that we find in Egypt
about the same elementary level we observed in contemporary
Mesopotamia. The areas of triangles, trapezoids, rectangles, etc.
are computed, and for the circle a rule is used which we can

transcribe as $A = \left(\dfrac{8}{9}d\right)^2$ if d denotes the diameter. Correspond-

ing formulae for the elementary volumes were known, including
a correct numerical computation for the volume of a truncated
pyramid. This, as well as the relatively accurate value 3.16 for π
resulting from the above formula, give Egyptian geometry a lead
over the corresponding arithmetical achievements. It has even
been claimed that the area of a hemisphere was correctly found
in an example of the Moscow papyrus, but the text admits also
of a much more primitive interpretation which is preferable.

A vivid description of the main topics of Egyptian mathematics
is given in a papyrus of the New Kingdom, written for school

purposes. It is a satirical letter in which an official ridicules a colleague. The section on mathematics runs as follows[1]): "Another topic. Behold, you come and fill me with your office. I will cause you to know how matters stand with you, when you say 'I am the scribe who issues commands to the army'.

"You are given a lake to dig. You come to me to inquire concerning the rations for the soldiers, and you say 'reckon it out'. You are deserting your office, and the task of teaching you to perform it falls on my shoulders.

"Come, that I may tell you more than you have said: I cause you to be abashed (?) when I disclose to you a command of your lord, you, who are his Royal Scribe, when you are led beneath the window (of the palace, where the king issues orders) in respect of any goodly (?) work, when the mountains are disgorging great monuments for Horus (the king), the lord of the Two Lands (Upper and Lower Egypt). For see, you are the clever scribe who is at the head of the troops. A (building-) ramp is to be constructed, 730 cubits long, 55 cubits wide, containing 120 compartments, and filled with reeds and beams; 60 cubits high at its summit, 30 cubits in the middle, with a batter of twice 15 cubits and its pavement 5 cubits[2]). The quantity of bricks needed for it is asked of the generals, and the scribes are all asked together, without one of them knowing anything. They all put their trust in you and say, 'You are the clever scribe, my friend! Decide for us quickly! Behold your name is famous; let none be found in this place to magnify the other thirty[3])! Do not let it be said of you that there are things which even you do not know. Answer us how many bricks are needed for it?

"See, its measurements (?) are before you. Each one of its compartments is 30 cubits and is 7 cubits broad."

On the whole, one can repeat here what we have already said for Babylonian geometry. Problems concerning areas or volumes do not constitute an independent field of mathematical research but are only one of many applications of numerical methods to practical problems. There is no essential difference between the determination of the acreage of a field in special measures and

[1]) From Erman, Egyptian Literature, p. 223 f.
[2]) These explanations are due to L. Borchardt. Cf. for a figure Quellen und Studien zur Geschichte der Mathematik, ser. B, vol. 1, p. 442.
[3]) Perhaps the frequently mentioned "College of the Thirty".

the distribution of beer to temple personnel according to different ratings. This is a state of affairs which holds to a large extent even in the Hellenistic period and far beyond it. In Arabic mathematics the "inheritance" problems play an important role, while similar examples are found already in Old-Babylonian texts. The geometrical writings of Heron, whether authentic or merely ascribed to him, contain whole chapters on units, weights, measurements, etc. Of course, since the Hellenistic period, even the writings of Heron and related documents show the influence of scientific Greek geometry. But, by and large, one has to distinguish two widely separate types of "Greek" mathematics. One is represented by the strictly logical approach of Euclid, Archimedes, Apollonius, etc.; the other group is only a part of general Hellenistic mathematics, the roots of which lie in the Babylonian and Egyptian procedures. The writings of Heron and Diophantus and works known only from fragments or from papyrus documents form part of this oriental tradition which can be followed into the Middle Ages both in the Arabic and in the western world. "Geometry" in the modern sense of this word owes very little to the modest amount of basic geometrical knowledge which was needed to satisfy practical ends. Mathematical geometry got one of its most important stimuli from the discovery of irrational numbers in the 4th or 5th century B.C. and remained rather stagnant from the second century B.C. onwards, except for those additions of spherical geometry and descriptive geometry which were introduced by their astronomical importance. On the other hand, geometrial theory had a negative effect on the algebraic and numerical methods which were part of the Oriental background of Hellenistic science. A real insight into the mutual relations between all these fields was not reached before modern times.

39. The role of Egyptian mathematics is probably best described as a retarding force upon numerical procedures. Egyptian astronomy had much less influence on the outside world for the very simple reason that it remained through all its history on an exceedingly crude level which had practically no relations to the rapidly growing mathematical astronomy of the Hellenistic age. Only in one point does the Egyptian tradition show a very beneficial influence, that is, in the use of the Egyptian calendar

by the Hellenistic astronomers. This calendar is, indeed, the only
intelligent calendar which ever existed in human history. A
year consists of 12 months of 30 days each and 5 additional days
at the end of each year. Though this calendar originated on purely
practical grounds, with no relation to astronomical problems, its
value for astronomical calculations was fully recognized by the
Hellenistic astronomers. Indeed a fixed time scale without any
intercalations whatsoever was exactly what was needed for
astronomical calculations. The strictly lunar calendar of the
Babylonians, with its dependence on all the complicated variations of the lunar motion, as well as the chaotic Greek calendars,
depending not only on the moon but also on local politics for its
intercalations, were obviously far inferior to the invariable
Egyptian calendar. It is a serious problem to determine the number
of days between two given Babylonian or Greek new year's days,
say 50 years apart. In Egypt this interval is simply 50 times
365. No wonder that the Egyptian calendar became the standard
astronomical system of reference which was kept alive through
the Middle Ages and was still used by Copernicus in his lunar
and planetary tables. Even in a civil calendar the Egyptian year
of 365 days was revived during the Middle Ages. The last Sasanian
king, Yazdigerd, based the reformed Persian calendar on this
year, shortly before the collapse of the Sasanian monarchy under
the impact of expanding Islam. Nevertheless the "Persian" years
of the Era Yazdigerd (beginning A.D. 632) survived and are often
referred to in Islamic and Byzantine astronomical treatises.

A second Egyptian contribution to astronomy is the division of
the day into 24 hours, though these "hours" were originally not
of even length but were dependent on the seasons. These "seasonal
hours", twelve for daylight, twelve for night, were replaced by
"equinoctial hours" of constant length only in theoretical works
of Hellenistic astronomy. Since at this period all astronomical
computations were carried out in the sexagesimal system, at least
as far as fractions are concerned, the equinoctial hours were
divided sexagesimally. Thus our present division of the day into
24 hours of 60 minutes each is the result of a Hellenistic modification of an Egyptian practice combined with Babylonian numerical
procedures.

Finally, we have to mention the "decans" (to use a Greek term) which have left no direct traces in modern astronomy. This is curious enough since the decans, as we shall see, are the actual reason for the 12-division of the night and hence, in the last analysis, of the 24-hour system. Again in Hellenistic times the Egyptian decans were brought into a fixed relation to the Babylonian zodiac which is attested in Egypt only since the reign of Alexander's successors. In this final version the 36 "decans" are simply the thirds of the zodiacal signs, each decan representing 10° of the ecliptic. Since the same period witnesses the rapid development of astrology, the decans assumed an important position in astrological lore and in kindred fields such as alchemy, the magic of stones and plants and their use in medicine. In this disguise the decans reached India, only to be returned in still more fantastic form to the Muslims and the West. Their final triumph lies in the frescoes of the Palazzo Schifanoria in Ferrara under Borso d'Este (about 1460).

In tracing back the history of the Egyptian decans we discover the interaction of the two main components of Egyptian time reckoning: the rising of Sirius as the harbinger of the inundation, and the simple scheme of the civil year of 12 months of three decades each.

39 a. Here is not the place to attempt a description of the history of the Egyptian calendar. Its basically non-astronomical character is underlined by the fact that the year is divided into three seasons of four months each, of purely agricultural significance. The only apparent astronomical concept is the heliacal rising of Sirius which, however, obtained its importance only by its closeness to the rising of the Nile, the main event in the life of Egypt. There existed, finally, a lunar calendar which regulated festivals in relation to the phases of the moon. As a matter of fact, as R. A. Parker has observed, there are different variants of lunar calendars to be distinguished, one of which was also eventually schematized and brought into a fixed relation to the schematic civil calendar with its twelve 30-day months and 5 epagomenal days.

When the decans first appear—on coffin lids of the Middle Kingdom—the civil calendar had long been established. To it is now set in relation a series of constellations, 36 in number, though with small variations in arrangement and limitations. Only two can be directly identified, namely Sirius and Orion. Some constellations cover more than one decan; on the other hand there are decans "preceding" or "following" a constellation, indicating groups of stars of lesser significance. We shall see that all these decans belong to a zone of the sky roughly parallel to and south of the ecliptic (cf. Fig. 3 b p. 87).

The astronomical representations on the coffin lids are, in all probability, poor replicas of ceilings in royal tombs or temples which were imitated in the modest coffins of minor people. These pictures represent the sky with the decanal constellations inscribed on them, arranged in their ten-day intervals throughout the year, forming 36 columns with 12 lines each for the 12 hours of the night. The name of a specific decan moves from column to column, each time one line higher. Thus there originated a diagonal pattern which is the motivation for the name "diagonal calendars" for these texts.

In fact, we have here not a calendar but a star clock. The user of this list would know the "hour" of night by the rising of the decan which is listed in the proper decade of the month. We shall now proceed to investigate the working of this type of "clock" more closely, at first from a modern point of view, then turning to historical considerations.

When we watch the stars rise over the eastern horizon, we see them appear night after night at the same spot on the horizon. But when we extend our observation into the period of twilight, fewer and fewer stars will be recognizable when they cross the horizon, and near sunrise all stars will have faded out altogether. Let us suppose that a certain star S was seen just rising at the beginning of dawn but vanished from sight within a very short time because of the rapid approach of daylight. We call this phenomenon the "heliacal rising" of S, using a term of Greek astronomy. Let us assume that we use this phenomenon as the indication of the end of "night" (meaning real darkness) and consider S as the star of "the last hour of night". One day later we may again say that the brief appearance of S indicates the end of night. We may continue in the same way for several days, but during this time a definite change takes place. The sun not only participates in the daily rotation of the sky from East to West, thus producing the change of night and daylight, but it also has a slow motion of its own relative to the stars in a direction opposite to the daily rotation.

This eastward motion of the sun (completed once in one year) delays the rising of the sun from day to day with respect to the rising of S. Consequently, the rising of S will be more and more clearly visible and it will take more and more time before S fades away in the light of the coming day. Obviously, after some lapse of time, it no longer makes sense to take S as the indicator of the last hour of night. But there are new stars which can take the place of S, and this procedure can be repeated all year long until the sun comes back to the region of S. Thus year after year S may serve for some days as the star of the last hour, to be replaced in regular order by other stars T, U, V, ...

It is this sequence of phenomena which led the Egyptians to measure the time of night by means of stars (or groups of nearby stars) which we now call the decans. In the above description, we left unanswered the tacit question: how long shall we wait until we replace S by T, T by U, etc.? Obviously, one could be very strict and choose daily a different star which is just in the phase of "heliacal rising". But this sort of impractical pedantry was not characteristic of those Egyptians, who intended to devise some method of indicating the times of office for the nightly service in the temples. They adjusted these times to their calendar. As the months were divided into decades, so were the services of the hour-stars. For 10 days, S indicated the last hour of night, then T was chosen for the next 10 days, and so forth. During each decade the end of night receded from dawn toward darkness, only to be pushed back toward dawn by the heliacal rising of the next "decan" as we shall now call the stars S, T, U, ...

So far we have only described the definition of the end of "night" or the last "hour". We have made one choice: we applied the decimal order of the civil calendar to these decanal hours. What follows is a necessary consequence of this vital decision.

We go back to the time of year when S serves as the decan of the last hour. Ten days later, T takes the place of S. By this time the rising of S is clearly visible in full darkness. Since the last hour is now indicated by T, we shall naturally say that the rising of S marks the hour next to the last. After another ten days, U represents the last hour, T the preceding hour, S the second-last hour, etc. Writing from right to left as the Egyptian texts do, we have thus obtained the following "diagonal" arrangement:

decade 3	decade 2	decade 1	
. . . .	. . . .		. . . .
. . . .	S		second to last hour
. . . .	T	S	next to last hour
. . . .	U	T	S	last hour

How long can this process be continued? To simplify matters, let us at first assume that a year had exactly 360 days or 36 decades. Then we need 36 decans before S can serve again as decan of the last hour. Our "star clock" will therefore be composed of 36 columns. The number of lines is a consequence of the following consideration. The rising of stars can only be seen at night. The maximum number of "hours" indicated by our decans is therefore equal to the number of decans which we can see rise in succession during one night. If we had complete darkness from the moment of sunset to the moment of sunrise, and if night and day were equal all year, one could always see exactly one-half of the celestial sphere rise during one night. Since 36 decans correspond to one complete circuit of the sky, 18 decans would be seen rising each night and our list of stars would lead to an 18-division of the night. In reality, however, the variation in the length of night as well as twilight influences this number considerably. A closer investigation shows that during summer, when Sirius rises heliacally, only 12 decans can be seen rising during darkness. Hence the decadic succession of the decans leads to a 12-division of the night. This, indeed, is the arrangement we find in the "diagonal calendars" on coffin lids of the period from about 1800 to 1200 B.C.

It is essential to recall that it was the decimal arrangement of the calendar which determined the spacing of the decans and thus the number of hours to be indicated by their rising each night. A finer division would have led to a greater number of hours, while longer intervals would have given fewer hours. The 12-division is therefore not an arbitrary choice of units, but the consequence of the decimal order of the civil calendar. The decimal basis of time reckoning appears in another form in the division of daylight. One of the inscriptions of the cenotaph of Seti I (about 1300 B.C.) shows a simple sun-dial and gives a description of its use. From this it follows that this instrument indicated ten "hours" between sunrise and sunset. To this, two more hours are added for morning and evening twilight respectively.

Thus we see that the Egyptian reckoning of hours was originally decimal for daylight, duodecimal for the time of darkness because of the decimal structure of the calendar, and leaving two more "hours" for twilight. The result is 24 "hours" of rather uneven length and uneven distribution between daylight and night. We do not know the details of the further development, but it can be shown that this primitive system was already obsolete when it was still depicted on the inscriptions of Seti I, giving way to a more even distribution of 24 hours into 12 hours of night and daylight each—a division which eventually led to the 24 "seasonal" hours of the Hellenistic period.

39 b. Not only must the independent division of darkness and daylight have soon become obsolete, but also the decanal clocks for the hours of night were bound to lose their usefulness in the course of a century or two. In our description, we assumed for the sake of simplicity a year of exactly 360 days' length. In this case, 36 decans would repeat their service periodically, following the diagonal pattern as described before. Actually, however, the Egyptian civil year contained 365 days. Since the 36 decans suffice only for the 360 days, an additional set of constellations is required to indicate the hours of darkness for the epagomenal days. All this was, in fact, taken into account by the inventors of the decanal hours, as can be demonstrated by the terminal section of the "diagonal calendars" on the coffin lids. What was not taken into consideration, however, is the fact that 365 days do not accurately measure the return of the sun to the same star, and consequently, a slow but relentless change in the relation between the heliacal rising of a decan and its date in the civil calendar takes place. Our texts show that rearrangements of the decanal order were attempted in order to counter the resulting disturbances. By the time of the New Kingdom, the usefulness of the decans as indicators of hours had ceased. An attempt to substitute the culmination of stars for their rising also did not last. But by this time the decans held a secure position as representatives of the decades of the year in the decoration of astronomical ceilings, as in the tomb of Senmut or in the cenotaph of Seti I. In this form they continued to exist until their association with the zodiac of the Hellenistic period revived them and made them powerful elements of astrological doctrine.

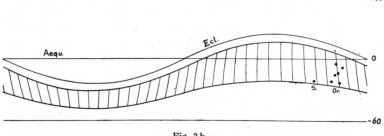

Fig. 3 b.

We still have to answer the question of the location in the sky of the decans when they were first invented as the indicators of the hours of night. From what has been said to this point, any sequence of stars or constellations whose risings were spaced at ten-day intervals could have been used. But additional information is available. We not only know that Sirius and Orion figured among the decans but that Sirius was, so to speak, the ideal prototype of all the other decans. Its heliacal rising ideally begins the year, just as the rising of the other decans are associated with the beginning of the parts of the year, the decades. The rising of Sirius occurs after an interval of about 70 days, in which the star remains invisible because of its closeness to the sun. Similarly, it was assumed that the same holds for all decans. The Demotic commentary to the inscriptions in the cenotaph of Seti I describes at length how one decan after another "dies", how it is "purified" in the embalming house of the nether world, to be reborn after 70 days of invisibility.

Such a mythological description cannot, of course, be taken as an exact astronomical condition for the duration of invisibility. But there can be little doubt that the decans were essentially to follow the cycle of Sirius. In other words, to serve as decan during the decade immediately after Sirius, a star must have been chosen that not only rose ten days later, but that also had a period of 70 days' invisibility. If these data were accurate and if the brightness of the stars in question were known, their positions could be determined accurately. This is not the case. Nevertheless, the deviation from a 70-day invisibility as well as the variation in brightness may be assumed to have remained within reasonable limits. This suffices to localize at least a zone within which constellations, which can serve as decans, must be located. The result is shown in Fig. 3 b[1]) which represents the belt of the decans in its relation to the ecliptic and equator and to Sirius and Orion. To attempt to go further in the determination of the decans is not only of very little interest but would necessarily imply ascribing to our texts an astronomical accuracy which they were never intended to have. But we have reached insight into a sound, however primitive, procedure of marking time at night by means of stars and are able to localize them in a definite region of the sky to which Sirius and Orion belong, not as exceptions, but as the leading members of the decanal constellations.

40. The coffins with the "diagonal calendars" belong roughly to the period from 2100 B.C. to 1800 B.C. From the New Kingdom, more elaborate monuments are preserved. One is the ceiling of the unfinished tomb of Senmut, the Vezir of Queen Hatshepsut; another is the ceiling of the cenotaph of King Seti I, about 1300 B.C. The tomb of Senmut contains lists of the decans, the representation of the deities of the hours, etc., and pictures of constellations of the northern hemisphere. On Pl. 10 part of this section is shown as copied by the expedition of the Metropolitan Museum in New York. The Hippopotamus and the Crocodiles etc. appear often again on similar pictures. Of special interest in the present drawing is the fact that it shows two stages of work. Below the sharply outlined drawings are visible faint lines which were drawn in blue on the ceiling. The man whose one arm is near a crocodile was missing on the original draft. The crocodile was not drawn in a skew position but horizontally. The traces are still visible at the feet and legs of the standing man. The lion, which

---

[1]) Fig. 3 b can be considered as an unrolling (or cylinder projection) of the celestial sphere between declinations $\pm$ 60° (or $\varphi = 30°$) with the equator as line of symmetry.

is now one register higher up, was lying parallel to the crocodile. The base line is still visible crossing the front shoulder of the skew crocodile. The head of the lion was at the man's shoulder, the lion's front paws are to the right of the man's belt. The previous inscription occupied the place of the final lion. These details are of interest because they demonstrate drastically that artistic principles determined the arrangement of astronomical ceiling decorations. Thus it is a hopeless task to try to find, on the sky, groups of stars whose arrangement might have been the same as the depicted constellations seem to require. Astronomical accuracy was nowhere seriously attempted in these documents.

In the tombs of Ramses VI, VII, and IX a new type of astronomical text appears. Here we find depicted observations which were made to determine the hours of the night throughout the year. For the first and for the 16th day of each month we see pictured a sitting man (Pl. 11) and above (or, better, behind) him a coordinate net with stars. The accompanying inscription mentions, for the beginning of the night and for each of the 12 hours of the night, a star and where it will be seen "over the left ear", "over the right ear", "over the left shoulder", or the "right shoulder", etc.

The horizontal lines in the coordinate network represent the hours, the vertical lines the positions. The stars are entered as named in the text—at least in principle, except for the innumerable errors which the craftsmen committed. Obviously we are dealing here with a much more refined method of time measurement than in the coffins of the Middle Kingdom. Nevertheless these texts were mechanically copied over much longer periods than they could possibly cover correctly. Much effort has been spent to identify these new lists of stars, often without the realization that the contents of the texts in a purely philological respect have not been safely established, because the available copies were made in the early days of Egyptology and often without consideration for variants in other copies and related texts. Only a new edition of this whole material can provide the necessary basis for such studies.

41. With the Ptolemaic period, Egyptian astronomy changes in aspect. A totally new element, the Greco-Babylonian zodiac, appears on the monuments. The ceilings of the Hellenistic temples erected and restored by Ptolemaic kings and Roman emperors, truly represent the chaotic mixture of astro-mythology and astrology of the Hellenistic period.

Beginning with the second century B.C., also astronomical (or, more accurately, calendaric) and astrological papyri appear, written in Greek or in Demotic or both. The earliest Demotic and Greek horoscopes were written close to the beginning of our era. Also proper "astronomical" texts written sometimes in Demotic, sometimes in Greek, have been found. We have planetary texts from the time of Augustus to Hadrian. In these, the dates when a planet enters a zodiacal sign are recorded. These texts are based on computations, not on observations, as is evident from the fact that entrances into a zodiacal sign are also noted when the planet is in conjunction with the sun, thus being invisible.

Another text of the Roman period, written in Demotic, undoubtedly represents an older Egyptian method, probably uninfluenced by Hellenism. We have mentioned before that lunar calendars played a role since early times side by side with the schematic civil calendar of the 365-day year. An inscription of the Middle Kingdom mentions "great" and "small" years, and we know now that the "great" years were civil years which contained 13 new-moon festivals in contrast to the ordinary "small" years with only 12 new moons. The way these intercalations were regulated, at least in the latest period, is shown by the Demotic text.

This Demotic text contains a simple periodic scheme which is based on the fact that 25 Egyptian civil years (which contain 9125 days) are very nearly equal to 309 mean lunar months. These 309 months are grouped by our text into 16 ordinary years of 12 lunar months, and 9 "great" years of 13 months. Ordinarily, two consecutive lunar months are given 59 days by our scheme, obviously because of the fact that one lunar month is close to $29\frac{1}{2}$ days long. But every 5th year the two last months are made 60 days long. This gives for the whole 25-year cycle the correct total of 9125 days.

In this way one had an exceedingly handy scheme of determining by means of a simple rule the dates of all lunar festivals in such a way that no grave error could develop for many centuries, though the single new moon or full moon could deviate by $\pm$ 2 days or perhaps even more. Yet the Egyptians were obviously satisfied with their scheme in the same way as the Jewish and Christian calendar of the Middle Ages confidently relied on a periodic scheme which showed equally serious deviations from the facts.

In summary, from the almost three millenia of Egyptian writing, the only texts which have come down to us and deal with a numerical prediction of astronomical phenomena belong to the Hellenistic or Roman period. None of the earlier astronomical documents contains mathematical elements; they are crude observational schemes, partly religious, partly practical in purpose. Ancient science was the product of a very few men; and these few happened not to be Egyptians.

BIBLIOGRAPHY TO CHAPTER IV

Our knowledge of Egyptian mathematics is primarily based on the following texts, all of which were written in the Middle Kingdom or the Hyksos period.

1. *Mathematical Papyrus Rhind* published first by Eisenlohr in 1877. Modern publication by T. E. Peet, London, 1923; additional material and photographs in Chace-Bull-Manning-Archibald, The Rhind Mathematical Papyrus, Oberlin, Ohio I 1927, II 1929.

2. *Moscow Mathematical Papyrus* published by W. Struve, Quellen und Studien zur Geschichte der Mathematik, ser. A. vol. 1 (1930).

3. *P. Berlin 6619.* Published by Schack-Schackenburg, Zeitschr. f. aegyptische Sprache 38 (1900) p. 135 ff. and 40 (1902) p. 65 f.

4. *P. Kahun.* Published by F. Ll. Griffith, Hieratic Papyri from Kahun and Gurob, London 1898 Pl. VIII and p. 15 ff.

5. *Leather roll British Museum 10250* by S. R. K. Glanville in J. Egyptian Archeology 13 (1927) p. 232 ff.

6. *Wooden tablets Cairo 25367 and 25368.* Recueil de travaux relatifs à la philologie et à l'archéologie égyptiennes 28 (1906) p. 62 ff. and Catalogue générale ... du Musée du Caire, Ostraca, 1901 Pl. 62–64 and p. 95 f.

For the late period, Demotic papyri should be added. One large Demotic text was found in Tunah el Gabal according to Ill. London News 104 (1939) p. 840 and Chronique d'Égypte 14 No. 28 (1939) p. 278. No information about this text could be obtained. Fragments of geometrical Demotic texts are in the Carlsberg Collection of the University of Copenhagen, to be published by A. Volten.

Greek papyri are very closely related to the Egyptian texts. For this material cf. André Deleage, Les cadastres antiques jusqu'à Dioclétien, Études de papyrologie 2 (1934) p. 73–228. K. Vogel, Beiträge zur griechischen Logistik I. Sitzungsber. d. Bayerischen Akademie der Wissensch., Math.-nat. Abt. 1936

p. 357–472. Cf. also Mitteilungen aus der Papyrussammlung der National-bibliothek in Wien, Griechische literarische Papyri I (1932) [Gerstinger and Vogel].

A great number of geometrical and arithmetical problems are found on papyri. A systematic investigation of this scattered material would be worth-while. I mention only especially large multiplication tables for fractions from the fourth century published as No. 146 in Michigan Papyri vol. III (1936). Many smaller tables are preserved both in Greek and in Demotic.

An excellent brief summary of Egyptian mathematics was given by T. E. Peet, Bull. John Rylands Library 15 (1931). A detailed study by the present author of the arithmetical methods is to be found in Quellen und Studien zur Geschichte der Mathematik, ser. B, vol. 1 (1930) p. 301–380, on geometry p. 413–451.

For a deeper understanding of the background which determined the character of Egyptian arithmetic, the study of the following works will be of great help: Lucien Lévy-Bruhl, Fonctions mentales dans les sociétés inférieures (1922); Heinrich Schaefer, Von aegyptischer Kunst (1919), and Kurt Sethe, Von Zahlen und Zahlworten (quoted p. 23).

General works on Egypt: J. H. Breasted, A History of Egypt, New York, Scribner. A. Erman, The Literature of the Ancient Egyptians; Poems, Narra-tives, and Manuals of Instruction, from the Third and Second Millenia B.C., New York, E. P. Dutton, 1927. Egyptian mathematics is described in O. Neu-gebauer, Vorgriechische Mathematik. Berlin, Springer, 1934 (quoted as "Vorlesungen"). There exists no modern work on Egyptian astronomy. An edition of all available Egyptian astronomical texts by R. A. Parker and the present author is in preparation. For literature concerning special problems see the notes given below to Nos. 39 ff.

NOTES AND REFERENCES TO CHAPTER IV

ad 36. A detailed analysis of the table for $\frac{2}{n}$ of the Papyrus Rhind is given in my book "Die Grundlagen der ägyptischen Bruchrechnung", Berlin, Springer, 1926. This theory is summarized in my "Vorlesungen" p. 137 ff. Modifications were proposed by van der Waerden "Die Entstehungsgeschichte der aegyp-tischen Bruchrechnung" in "Quellen und Studien zur Geschichte der Mathematik" ser. B vol. 4 (1937) p. 359–382.

ad 37. As stated in the text we find already in the New Kingdom an exception to the rules of the Papyrus Rhind for the duplication of unit fractions. William C. Hayes, Ostraca and Name Stones from the Tomb of Sen-Mut (No. 71) at Thebes, The Metropolitan Museum of Art, Egyptian Expedition [Publications No. 15] New York 1942, published an ostracon (No. 153) which contains the following computation[1])

[1]) The restoration of the original problem given by Hayes seems to me very doubtful. In the first line one can read safely only 3 14 2 21 and I see no reason for restoring "cubit, palm(?)" at the beginning. The four fractions obviously form two pairs but I do not understand their relation to the subsequent operations.

1	$\bar{7}$			(black)
	3			(red)
2	$\bar{6}$	$\overline{14}$	$\overline{21}$	(black)
3	$\bar{2}$ 1 $\bar{2}$	1		(red)
4	$\bar{2}$	$\overline{14}$		(black)
10	$\bar{2}$ 1 $\bar{2}$			(red)

where the numbers below the main lines are written in red. Obviously we are dealing here with a multiplication of $\bar{7}$. The standard procedure would be

$$
\begin{array}{ll}
1 & \bar{7} \\
2 & \bar{4}\ \overline{28} \\
4 & \bar{2}\ \overline{14}
\end{array}
$$

Thus we see that the ostracon uses a different (and more complicated) expression for twice $\bar{7}$. The analysis of this decomposition is useful for the understanding of the method which we have described in the text. The standard decomposition would operate with the natural fraction $\bar{4}$ and determine the fraction which remains for $2 - \bar{4}$ of $\bar{7}$. The result is $\bar{4}$.

In the case of the ostracon we find "auxiliary numbers" written in red below the fractions. Under $\overline{21}$ we find 1. This means that $\overline{21}$ is introduced as a new unit; consequently we find 3 below $\bar{7}$. This shows us that we are dealing, not with the natural fraction $\bar{4}$ of the sequence $\bar{2}$, $\bar{4}$, ..., but with $\bar{3}$ which belongs to the sequence $\bar{3}$, $\bar{3}$, $\bar{6}$, ... Hence we obtain now $\overline{21}$ as one term and must find the remainder which is obtained by multiplying $\bar{7}$ by $2 - \bar{3} = 1\ \bar{\bar{3}}$. We know already that $\bar{\bar{3}} = \bar{2}\ \bar{6}$. Thus we have to multiply $\bar{7}$ by $1\ \bar{2}\ \bar{6}$. Here again auxiliary numbers must be introduced, counting $\bar{6}$ as 1, which will lead to

$$
\begin{array}{rll}
/ & 1 & 6 \\
& \bar{2} & 3 \\
/ & \bar{6} & 1
\end{array}
$$

If we take here the first and last term we have 7 new units. Thus we see that $1\ \bar{6}$ of $\bar{7}$ is $\bar{6}$. There remain $\bar{2}$ of $\bar{7}$ which is $\overline{14}$. Thus we have found for the remainder $\bar{6}\ \overline{14}$ and for the whole of twice $\bar{7}$ the form

$$
\bar{6}\quad \overline{14}\quad \overline{21}.
$$

The above computation shows how important it is to begin with the proper natural fraction. The use of $\bar{4}$ leads to a two-term expression, whereas the use of $\bar{3}$ forced us into a three-term decomposition. I am sure that the Egyptians never saw behind the underlying reason of divisibility but simply operated by trial and error. The reader might find the above explanation exceedingly clumsy and hypothetical. Only a systematic study of many available examples can give the necessary experience so that one becomes really familiar with this type of arithmetical rules. A useful illustration, however, is a group of problems from the mathematical Papyrus Rhind which I have analyzed in detail in my "Vor-

lesungen", p. 139 ff. Fortunately, these examples also deal with $\overline{7}$ and its sub-
fractions, and in the majority of cases all the auxiliary numbers which help to
"multiply" fractions are preserved.

ad 39. For Egyptian time reckoning see K. Sethe, Die Zeitrechnung der
alten Aegypter im Verhältnis zu der der andern Völker. Nachr. d. k. Gesellschaft
d. Wissensch. zu Göttingen, Phil.-hist. Kl. 1919 and 1920. Also L. Borchardt,
Die altägyptische Zeitmessung, Berlin, De Gruyter, 1920. The Egyptian lunar
calendars are discussed by R. A. Parker, The Calendars of Ancient Egypt,
University of Chicago Press, 1950.

For the later history of the decans see W. Gundel, Dekane und Dekanstern-
bilder. Studien d. Bibliothek Warburg 19 (1936).

ad 39 a. For the origin of the Egyptian calendar cf. O. Neugebauer, Die
Bedeutungslosigkeit der 'Sothis-periode' für die älteste ägyptische Chronologie,
Acta Orientalia 17 (1938) and The Origin of the Egyptian Calendar, J. Near
Eastern Studies 1 (1942). Also H. E. Winlock, The Origin of the Ancient
Egyptian Calendar, Proc. Am. Philos. Soc. 83 (1940), and R. A. Parker in
the book quoted in the preceding section.

The "diagonal calenders" were first discussed by A. Pogo, Isis 17 and 18
(1932) and Osiris 1 (1936). The location of the decans from their period of
invisibility was given by O. Neugebauer in "Vistas in Astronomy" (ed. by
Arthur Beer) vol. I p. 47–51 (London, 1955). For the tomb of Senmut see
A. Pogo, Isis 14 (1930).

The Seti Cenotaph was published by H. Frankfort, in Memoir 39 of the
Egypt Exploration Society (2 vols.) London 1933. For the discussion of the
astronomical ceiling cf. H. O. Lange–O. Neugebauer, quoted p. 69 ad No. 29.

A very puzzling text from a Ramesside papyrus (concerning lucky and
unlucky days) was published by J. Černý, Annales du Service des Antiquités
de l'Égypte 43 (1943) p. 179 f. There we find a scheme for the length of daylight
and night from month to month, varying linearly between a minimum of 6
hours and a maximum of 18 hours. Another scheme for the variation of
daylight is discussed by J. J. Clère. Un texte astronomique de Tanis, Kêmi
10 (1949) p. 3–27.

ad 41. For the planetary tables cf. O. Neugebauer, Trans. Am. Philos.
Soc., N.S. 32 (1942) with additions in Knudtzon-Neugebauer, Zwei astro-
nomische Texte, Bull. de la soc. royale des lettres de Lund 1946–1947 p. 77 ff.
Discussion by van der Waerden, Egyptian 'Eternal Tables', Koninkl. Nederl.
Akad. van Wetensch., Proc. 50 (1947) p. 536 ff. and p. 782 ff.

The dating of four of these planetary tables was related to a peculiar accident
which is worth mentioning as an example of how the most unlikely combinations
may occur and mislead us in our conclusions. The four tables under discussion
are inscribed on wooden tablets which were originally bound together like the
pages of a book by means of strings which were strung through holes in one side
of the wooden frame (cf. Pl. 13 which shows Tablet II). These tablets were first
published in 1856 by Brugsch, one of the great pioneers of Egyptology. Each
tablet mentions regnal years and it was natural to arrange them accordingly,
because these years formed a complete sequence as follows

Tablet I   year 9 to 15
Tablet II   year 16 to 19 and 1 to 3
Tablet III year 4 to 10
Tablet IV   year 11 to 17

Because these texts were obviously written in the Roman period, Brugsch concluded that the first ruler must be Trajan, whose reign lasted 19 years and whose successor was Hadrian, whose reign lasted longer than 17 years. These conclusions proved to be correct, however, only for tablets I, II, and IV. Checking of the astronomical data showed immediately that No. III could not be the continuation of No. II nor could it be the predecessor of No. IV[1]). Indeed it was easy to show that the years "4 to 10" were not the years of Hadrian but of Vespasian, 30 years before Trajan. Hence we know that by mere accident Tablet III seems to fit between II and IV. Similar cases may occur, more often than we think, in historical research but escape discovery simply because the rigorous astronomical check is not applicable.

The 25-year cycle was discovered in the Demotic papyrus No. 9 of the Carlsberg collection, published by O. Neugebauer and A. Volten in Quellen und Studien zur Geschichte d. Mathematik, ser. B, vol. 4 (1938). This 25-year cycle was well known and often used in Hellenistic astronomy. Ptolemy, e. g., arranges his tables of syzygies according to it (Almagest VI, 3).

One must not misinterpret the expression "25-year cycle" as a parallel to the previously mentioned "19-year cycle" (or "Metonic cycle"; cf. p. 7). In the first case the 25 years are Egyptian calendar years of exactly 365 days each. In the second case the years are tropical years, i. e., time intervals which are astronomically defined and which involve fractions of days. The first cycle comprises 309 mean lunar months at the end of which the same Egyptian civil day appears again as the date of a new moon or full moon. In the second cycle 235 mean lunar months bring the same lunar phase back to the same season, but it depends on the local calendar whether or not this restores also the calendar date. Because the Greek astronomers operated consistently with the Egyptian calendar in their tables, the 25 year cycle was by far the most convenient cycle to use.

From the enormous wealth of written documents from ancient Egypt we have only one doubtful reference to a partial solar eclipse of 610 B.C. — assuming that this is the correct interpretation of the text (cf. W. Erichsen in Akad. d. Wiss. u. Lit. Mainz, Abh. Geistes- u. Soz. Wiss. 1956, No. 2.[2]) Not a single Egyptian observation is quoted in the Almagest, although Ptolemy gives extensive references to earlier observations on which his theory is based. There exists one Coptic eclipse record of 601 A.D.(!), first identified by Krall and Ginzel

---

[1]) This was correctly realized by William Ellis, Memoirs Roy. Astron. Soc. 25 (1857) p. 112 but, strangely enough, Ellis did not determine the correct date of No. III.

[2]) The alleged eclipse report of Osorkon (9th cent. B.C.) does not concern an actual eclipse, as was shown by R. Caminos in Analecta Orientalia 37 (1958) p. 88 ff.

(S. B. Akad. d. Wiss. Wien, math.-nat. Cl. 88, 2 [1883] p. 655) and again by E. B. Allen, J. Am. Oriental Soc. 67 (1947) p. 267.

Appendix. The reader may have missed a reference to the astronomical or mathematical significance of the pyramids. Indeed, a whole literature has been built up around the "mysteries" of these structures, or at least one of them, the pyramid of Khufu (or "Cheops"). Important mathematical constants, e. g., an accurate value of π, and deep astronomical knowledge are supposed to be expressed in the dimensions and orientation of this building. These theories contradict flatly all sound knowledge obtained by archeology and by Egyptological research about the history and purpose of the pyramids. The reader who wants to see an excellent account of these facts should consult the paper by Noel F. Wheeler, Pyramids and their Purpose, Antiquity 9 (1935) p. 5–21, 161–189, 292–304 and L. Borchardt, Gegen die Zahlenmystik an der grossen Pyramide bei Gise, Berlin 1922.

For the very complex historical and archaeological problems connected with the pyramids, cf., e. g., J. P. Lauer, Le problème des pyramides d'Égypte, Paris 1948, and I. E. S. Edwards, The Pyramids of Egypt, Penguin Books, 1952. How little one knows about the significance of the arrangement of rooms and corridors in the interior is particularly evident in the case of the "Bent Pyramid" at Dahshur: cf. A. Fakhry's recent excavation reports in Annales der Service des Antiquités de l'Égypte 51 (1954) p. 509 ff and 52 (1955) p. 563 ff.

CHAPTER V

Babylonian Astronomy.

42. There is scarcely another chapter in the history of science
where an equally deep gap exists between the generally accepted
description of a period and the results which have slowly emerged
from a detailed investigation of the source material. This discrepancy has its roots as far back as the Hellenistic tradition about
the "Babylonians" or "Chaldeans" who are innumerably many
times mentioned in ancient writings, especially in the astrological
literature. Thus magic, number mysticism, astrology are ordinarily
considered to be the guiding forces in Babylonian science. As far
as mathematics is concerned, these ideas have had to be most
drastically revised since the decipherment of mathematical texts
in 1929. But for more than 70 years the same sort of revision
resulted from the discoveries of Epping and Kugler in Babylonian
astronomy. Thanks to the work of these scholars, it very soon
became evident that mathematical theory played the major role
in Babylonian astronomy as compared with the very modest
role of observations, whose legendary accuracy also appeared
more and more to be only a myth. Simultaneously the age of
Babylonian astronomy had to be redefined. Early Mesopotamian
astronomy appeared to be crude and merely qualitative, quite
similar to contemporary Egyptian astronomy. At best since the
Assyrian period, a turn toward mathematical description becomes
visible and only the last three centuries B.C. furnished us with
texts based on a consistent mathematical theory of lunar and
planetary motion. The latest astronomical text has been identified
recently by Sachs and Schaumberger, with the date of 75 A.D.
These late theories, on the other hand, proved to be of the highest
level, fully comparable to the corresponding Greek systems and
of truly mathematical character. Simultaneously it had to be

admitted that we know next to nothing about the details of horoscopic astrology in Mesopotamia in sharpest contrast to the overwhelming abundance of astrological documents from Hellenistic Egypt and the Roman and Byzantine period.

Finally it has been repeatedly remarked by competent observers that the almost proverbial brilliance of the Babylonian sky is more a literary cliché than an actual fact. The closeness of the desert with its sand storms frequently obscures the horizon. This is the more essential as the majority of problems in which the Babylonian astronomers were interested are phenomena close to the horizon. The lunar calendar requires observation of the first visibility of the new crescent in the western horizon. The last visibility of the moon happens at the eastern horizon. Disappearance and reappearance of the planets are phenomena close to the horizon and it seems that also "opposition" of a planet was defined as rising or setting at sunset and sunrise respectively. Only eclipses and occultations will usually be observable under favorable conditions. It is certainly the result of this situation that Ptolemy states that practically complete lists of eclipses are available since the reign of Nabonassar (747 B.C.) while he complains about the lack of reliable planetary observations. He remarks that the old observations were made with little competence, because they were concerned with appearances and disappearances and with stationary points, phenomena which by their very nature are very difficult to observe. It is worth noting that this precise description of the planetary observations of the Babylonians by a competent astronomer had almost no effect on the current evaluation of Babylonian astronomy while the vague but abundant references of the Hellenistic astrologers to Chaldean wisdom completely dominated the picture which later centuries developed of Chaldean astronomy.

43. Our description of Babylonian astronomy will be rather incomplete. The historical development will be given in bare outline. As in the case of Egypt. a detailed discussion of the few preserved early texts would require not only too much room but would also unduly exaggerate their historical importance. For the late period, however, the opposite situation prevails. A great number of texts exist from the Seleucid epoch and only a very extensive and highly technical discussion could do justice to the

mathematical and astronomical achievements of this period. Obviously it is impossible to do this in the present frame.

We begin our survey with a short description of the earlier development. Then, before entering the discussion of the Seleucid period, a few remarks about our source material and its provenance will be necessary. Our description of the mathematical astronomy of the Hellenistic period we shall start with an outline of the theory of the solar and lunar motion because this theory is undoubtedly the most characteristic section of this whole development. The planetary theory will then be summarized so far as essentially new ideas which go beyond the methods already known from the luni-solar problems are apparent. We shall, finally, touch upon the few facts which are known about the milieu in which these texts originated.

44. To begin our historical sketch with a negative statement, we can say that nothing is known about a Sumerian astronomy. Mythological concepts which involve the heavens, deification of Sun, Moon, or Venus cannot be called astronomy if one is not willing to count as hydrodynamics the existence of belief in a storm deity or the personification of a river. Also the denomination of conspicuous stars or constellations does not constitute an astronomical science.

One of the earliest documents with definitely astronomical trend is a tablet in the Hilprecht Collection in Jena, Germany. The text was probably written in the Cassite period but copied from an original composition which was older. Its formulation is quite similar to a familiar type of Old-Babylonian mathematical texts. The document begins with a list of numbers and names which might be interpreted as follows: "19 from the Moon to the Pleiades; 17 from the Pleiades to Orion; 14 from Orion to Sirius", and so on for eight stars or constellation, ending with the statement that the total (of what?) is 120 "miles" and the question "how much is one god (i. e., star) beyond the other god?" Then begins the "procedure", exactly as in a mathematical text. It consists in dividing each of the given numbers by their sum, which is 1,21, well known to us as the last entry in the standard table of reciprocals (cf. p. 32). Each of these results is converted into "miles" and lower units of distance and explained as the distance from one of the previously enumerated celestial objects to the next. The text ends with the customary "such is the procedure" and the names of the scribe who copied the text and the one who verified the copy.

This text and a few similar fragments seem to indicate something like a universe of 8 different spheres, beginning with the sphere of the moon. This model obviously belongs to a rather early stage of development of which no traces have been found preserved in the later mathematical astronomy, which seems to operate without any underlying physical model. It must be emphasized, however, that the interpretation of this Nippur text and its parallels is far from secure.

Another group of probably contemporary texts represents a division of the sky into three zones of 12 sectors each. Each zone contains the names of constellations and planets and simple numbers in arithmetic progression like 1 1,10 1,20 etc. up to 2 and down again 1,50 1,40 until 1. This is probably the earliest occurrence of an arithmetical scheme which was later developed into an important tool for the description of periodic phenomena, the so-called zigzag functions. In the present case, the numbers are so simple and so obviously schematic that many different interpretations which explain them equally well or equally badly can be proposed.

There is another class of early documents which deserves mention because it contains the earliest records of actual observations in Mesopotamia. For several years of the reign of Ammisaduqa the appearances and disappearances of Venus were recorded. Because the dates are given in the contemporary lunar calendar, these documents have become an important element for the determination of the chronology of the Hammurapi period. From the purely astronomical viewpoint these observations are not very remarkable. They were probably made in order to provide empirical material for omina; important events in the life of the state were correlated with important celestial phenomena, exactly as specific appearances on the livers of sacrificial sheep were carefully recorded in the omen literature. Thus we find already in this early period the first signs of a development which would lead centuries later to judicial astrology and, finally, to the personal or horoscopic astrology of the Hellenistic age.

It is difficult to say when and how the celestial omens developed. The existing texts are part of large series of texts, the most important one called "Enūma Anu Enlil" from its initial sentence, similar to papal bullae in the Middle Ages. This series contained at least 70 numbered tablets with a total of about 7000 omens. The canonization of this enormous mass of omens must have extended over several centuries and reached its final form perhaps around 1000 B.C.

Historically much more interesting than this mass of purely descriptive omens are two texts which were called "mul apin". The earliest preserved copies are dated around 700 B.C., but they are undoubtedly based on older material. They contain a summary of the astronomical knowledge of their time. The first tablet is mostly concerned with the fixed stars which are arranged in three "roads", the middle one being an equatorial belt of about 30° width. The second tablet concerns the planets, the moon, the seasons, lengths of shadow, and related problems. These texts are incompletely published and even the published parts are full of difficulties in detail So much, however, is clear: we find here a discussion of elementary astronomical concepts, still quite descriptive in character but on a purely rational basis. The data on risings and settings, though still in a rather schematic form, are our main basis for the identification of the Babylonian constellations.

Around 700 B.C., under the Assyrian empire, we meet with systematic observational reports of astronomers to the court. Obviously the celestial omens have now reached primary importance. In these reports no clear distinction is yet made between astronomical and meteorological phenomena. Clouds and halos are on equal footing with eclipses. Nevertheless, it had been already recognized that solar eclipses are only possible at the end of a month (new moon), lunar eclipses at the middle. The classical rule that lunar eclipses are separated from one another by six months, or occasionally by five months only, might well have been known in this period. We should recall here Ptolemy's statement that eclipse records were available to him from the time of Nabonassar (747 B.C.) onwards.

It is very difficult to say when this phase developed into a systematic mathematical theory. It is my guess that this happened

comparatively rapidly and not before 500 B.C. Up to about
480 B.C., the intercalations of the lunar calendar show no regularity whatsoever. One century later, however, the rule of 7 intercalations in 19 years at fixed intervals seems to be in use, and
remains from now on the basis of all the lunar calendars which
were derived from the Babylonian scheme, including the lunar
calendar of the Middle Ages discussed in the first chapter.

A luni-solar intercalation rule presupposes the recognition of
a relation which indicates that m lunar months are equal in
length to n solar years. In the specific case of the 19-year cycle
$m = 235$ and $n = 19$. In the preceding period a "year" was an
interval of sometimes 12 or sometimes 13 months, where probably
the state of the harvest decided the need for a 13th month. The
existence of a cycle, however, proves that a more precise astronomical definition of "year" was adopted. We cannot give accurate
data about the mean length of such a year or how it was determined. There are good reasons, however, which point to an observation of the summer solstice as the point of comparison. At
any rate, it is the summer solstices which are systematically computed, whereas the equinoxes and the winter solstices are simply
placed at equal intervals. Because much more accurate methods
were known in the Seleucid period, it is plausible to assume that
the scheme of the 19-year cycle represents a slightly earlier phase
of development.

We shall see that period relations of the above-mentioned type
form the very backbone of Babylonian mathematical astronomy;
these are relations which state that s intervals of one kind equal
t intervals of another kind. Mathematical astronomy is fully
developed at about 300 B.C. at the latest. The 19-year intercalation cycle is certainly one of the most important steps preceding the later astronomical methods, that is to say, later than
about 450 B.C. Roughly to the same period, probably the fourth
century, belongs also the invention of the zodiac. The constellations
which lent their names to the zodiacal signs are, of course, much
older. But it was only for mathematical reasons that a definite
great circle which measured the progress of the sun and the planets
with respect to exactly 30°-long sections was introduced. Indeed,
the zodiac was hardly ever more than a mathematical idealization
needed and used exclusively for computing purposes. Actual

positions in the sky were expressed until the end of cuneiform writing with reference to well known bright stars. This primitive system was still in use in Greek horoscopes of the Roman period, exactly as in Babylonian texts, side by side with the determination of positions by degrees and zodiacal signs.

We may now enumerate the tools which were available at the end of the "prehistory" of Babylonian astronomy which extends from about 1800 B.C. to about 400 B.C. The zodiac of 12 times 30 degrees as reference system for solar and planetary motion. A fixed luni-solar calendar and probably some of the basic period relations for the moon and the planets. An empirical insight into the main sequences of planetary and lunar phenomena and the variation of the length of daylight and night. The use of arithmetic progressions to describe periodically variable quantities. And, above all, a complete mastery of numerical methods which could immediately be applied to astronomical problems. The utilization of these possibilities marks indeed the crucial step.

45. The next section will bring us to the discussion of the completed system of the Babylonian lunar theory. This discussion is based on texts whose significance was first recognized by Fathers Epping and Strassmaier. In 1881 there appeared in a Catholic theological periodical, the "Stimmen aus Maria Laach", an article "Zur Entzifferung der astronomischen Tafeln der Chaldäer" by J. Epping of Quito, Equador, with an introduction by J. N. Strassmaier in London. This paper concerns a fascinating report of the first decipherment of astronomical tablets which then were arriving in London in ever increasing numbers. The two authors were fully conscious of the importance of their discoveries. Indeed, this first paper contained the correct determination of the zero point of the Seleucid Era and that of the Parthian Era, thus providing for the first time a solid chronological basis for the history of Mesopotamia after Alexander the Great. But much more was done for the understanding of Babylonian astronomy itself. Suddenly it became clear that arithmetical progressions were skillfully utilized for the prediction of lunar phenomena, with an accuracy of a few minutes. The names of the planets and of zodiacal constellations were correctly determined and the road opened for the translation of astronomical records. On ten

pages Epping described discoveries which were to inaugurate a
new epoch in the history of science.

Eight years later Epping published a small book, entitled
"Astronomisches aus Babylon" in the supplements to the "Stimmen aus Maria Laach". Here one finds an account of the
leading ideas of the Babylonian theory of the moon as well as
a detailed discussion of planetary and lunar almanacs. Epping
died in 1894. The period of initial discoveries came to a conclusion
in the monumental works by Father Kugler, published between
1900 and 1924.

The texts on which Epping's and Kugler's work are based
come exclusively from the British Museum. For many years
Strassmaier copied there thousands of tablets, the majority of
which belong to the latest periods of Babylonian history. These
copies were collected in notebooks, of which one page is reproduced as a characteristic sample on Pl. 14. It was on Strassmaier's initiative that Epping began the study of Strassmaier's
transcriptions of astronomical texts. When the decipherment
proved to be successful, Strassmaier excerpted, from his notebooks, astronomical texts on special sheets, often adding explanatory remarks. These sheets were then sent to Epping for
final investigation, and after Epping's death, to Kugler. Kugler's
successor, Father Schaumberger, and I myself got the main
portion of our texts from Strassmaier's copies which he had
entered in his voluminous notebooks during the 1880's and
1890's. Not a single one of these texts was ever published in the
official publications of the British Museum; and no information
whatsoever is available concerning similar tablets which the
British Museum may have acquired after Strassmaier ceased
copying. Without Strassmaier, Epping, and Kugler, the few other
astronomical texts so far published would probably have been
laid aside in other museums too. It is very likely that no trace
of this enormously rich material would have penetrated to the
outside world, and Babylonian astronomy would still appear to
us in the light of a few texts from the earliest periods and of the
omens of Enūma-Anu-Enlil. A few numbers will illustrate this
situation. From Strassmaier's notebooks and from Kugler's publications about 240 astronomical texts and fragments were recovered, all of which were probably found in one archive in Baby-

lon. From the inventory numbers of the British Museum, one can conclude that these texts reached the museum between November, 1876 and July, 1882. During these six years the number of tablets increased from over 32,000 to more than 46,000 and one could expect that many hundreds of astronomical texts would be among these masses of texts. Indeed, in 1953 it became known that T. G. Pinches, before 1900, had copied some 1300 pieces of astronomical texts. This material was then put at the disposal of A. J. Sachs who published it with the addition of many related pieces in 1955. Thus we have now the major part of one ancient archive at our disposal, as far as it had reached the British Museum.

46. The mathematical astronomical texts fall into two major groups: "procedure texts" and "ephemerides". The texts of the first class contain the rules for the computation of the "ephemerides", which, in turn, are similar to a modern "nautical almanac", giving for a specific year (or for some specific sequence of years) the lunar or planetary positions at regular intervals. If the "procedure texts" were complete and if we fully understood their technical language, they might suffice for the actual computation of the ephemerides. In fact, however, none of these assumptions is satisfied. The preserved texts are badly damaged or totally missing for many of the steps; their terminology is far from clear, at least to us; and it might be justly asked if even a complete set of procedure texts would not have required supplementary oral explanation before it could be used for actually computing an ephemeris. Consequently the ephemerides themselves form the major basis for our researches, and the procedure texts often play the role of very welcome testing material for the rules which we finally abstract from the completed ephemerides. In the subsequent discussion we shall, however, make no sharp distinction between these two groups of sources and we shall act as if we had explicit rules at our disposal, though they are often actually only obtained from a very complex interplay between related fragments of both classes of texts.

The number of available astronomical tablets from the Seleucid period is not at all large. I know of less than 250 ephemerides, more than half of which are lunar, the rest planetary. The number of procedure texts is about 70, the majority of which are only small fragments. Thus our knowledge of Babylonian mathematical

astronomy is based on about 300 tablets. To this number can
now be added the vast mass of some 1000 non-mathematical
astronomical texts from Pinches-Sachs; it will take many years
of patient work before the conclusions can be drawn from this
great variety of new sources for the earlier development of
Babylonian astronomy in all its theoretical and practical aspects.

47. The fundamental problem of the Babylonian lunar theory
is determined by the calendar. So far as we know, the Babylonian
calendar was at all periods truly lunar, that is to say, the "month"
began with the evening when the new crescent was for the first
time again visible shortly after sunset. Consequently the Babylonian "day" also begins in the evening and the "first" of a
month is the day of the first visibility. In this way the beginning of a month is made dependent upon a natural phenomenon which is amenable to direct observation. This is certainly
a very simple and natural definition, as simple as the concurrent definition of the "day" as the time from one sunset to
the next. But as is often the case, a "natural" definition leads
to exceedingly complicated problems as soon as one wishes to
predict its consequences. This fact is drastically demonstrated
in the case of the lunar months. A very short analysis will illustrate the intrinsic difficulties.

A "lunar month" obviously contains an integer number of
days. How many? A rough estimate is easily obtainable. No two
consecutive reappearances of the new crescent after a short
period of invisibility of the moon are ever separated by more
than 30 days or by less than 29 days. Thus immediately the
main problem arises: when is a month 30 days long, when 29?
To answer this problem we must obtain an estimate not only of
the lunar motion, but also of the motion of the sun. In one year
of, roughly, 365 days, the sun moves once around us; that is
to say, after this time the sun again comes back to the same
star, having completed a great circle of 360°. Thus the solar
motion per day is close to 1° and therefore close to 30° in one
month. The time from one new crescent to the next is obviously
about equal to the time from invisibility to invisibility. But the
moon is invisible because it is close to the sun. Thus a month
is measured by the time from one "conjunction" of the moon
with the sun to the next. During this time the sun traveled about

30°; the moon, however, traveled not only 30°, but completed
one additional whole rotation of 360°. Hence 390° are covered
in about 30 days; this shows us that the moon must cover about
13° per day.

Now the real difficulties begin. In order to make the first
crescent visible the sun must be sufficiently deep below the
horizon to make the moon visible shortly before it is setting
(Fig. 4). The evening before, the moon was still too close to the
sun to be seen. Hence it is necessary to determine the distance
from the sun to the moon which is required to obtain visibility.

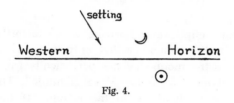

Fig. 4.

The distance between them depends on the relative velocity of the
two bodies. We have found that the moon moves 13° per day,
the sun 1° per day; thus the distance in question, the so-called
"elongation", increases about 12° per day. But this estimate is
no longer accurate enough to answer the question as to the
moment when the proper elongation is reached. Neither the sun
nor the moon moves with constant speed. Thus the daily elongation might vary between about 10° and 14° per day. This shows
that our problem involves the detailed knowledge of the variation
of both solar and lunar velocity.

But even if we had insight into the variable velocity of both
bodies the visibility problem would not be solved. For a given
place, all stars set and rise at fixed angles which are determined
by the inclination of the equator and the horizon. The relative
motion which we were discussing before is a motion in the ecliptic,
which makes an angle of about 24° with the equator. Consequently
we must know the variations of the angles between ecliptic and
horizon. For Babylon we find a variation from almost 30° to
almost 80° (Fig. 5). Thus the same elongation produces totally
different visibility conditions at different times of the year.

Let us assume that also the problem of the variation of the

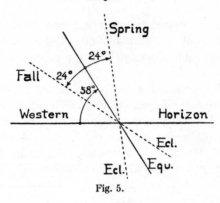

Fig. 5.

angles between ecliptic and horizon is satisfactorily answered. Then we must still remember that only the sun travels in the ecliptic whereas the moon deviates periodically from it between the limits of about $+5°$ and $-5°$ in "latitude". This deviation is measured perpendicularly to the ecliptic. If the ecliptic is almost vertical to the horizon (as is the case in spring), then the latitude has relatively little effect upon the visibility (cf. Fig. 6). In the fall, however, the full effect of the latitude is felt in bringing

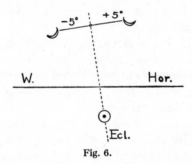

Fig. 6.

the moon nearer to, or farther away from, the horizon (Fig. 7). Thus we need also the knowledge of the variation of the latitude of the moon.

All these effects act independently of each other and cause quite irregular patterns in the variation of the length of lunar months. It is one of the most brilliant achievements in the exact sciences of antiquity to have recognized the independence of

ed to clarify the single steps but also to illustrate a pre-
y made remark to the effect that at no point of this theory
he traces of a specific geometrical model visible.

. The main tool for the computation of the ephemerides is
hmetic progressions, increasing and decreasing with con-
nt difference between fixed limits. As an example we offer the
lowing excerpt from the first three columns of an ephemeris for
e year 179 Seleucid Era, i. e. 133/132 B.C.:

	XII₂	28,55,57,58	22, 8,18,16	♈
2,59	I	28,37,57,58	20,46,16,14	♉
	II	28,19,57,58	19, 6,14,12	♊
	III	28,19,21,22	17,25,35,34	♋
	IV	28,37,21,22	16, 2,56,56	♌
	V	28,55,21,22	14,58,18,18	♍
	VI	29,13,21,22	14,11,39,40	♎
	VII	29,31,21,22	13,43, 1, 2	♏
	VIII	29,49,21,22	13,32,22,24	♐
	IX	29,56,36,38	13,28,59, 2	♑
	X	29,38,36,38	13, 7,35,40	♒
	XI	29,20,36,38	12,28,12,18	♓
	XII	29, 2,36,38	11,30,48,56	♈

In the first column we have dates, beginning with an intercalary (13th) month, called XII₂; then follows the year 2,59 of
the Seleucid Era and all the months from I to XII of this year.
These dates have a more accurate meaning. The first XII₂ does
not mean the whole month XII₂ but the moment of the mean
conjunction which falls at the end of this month. Similarly, each
subsequent month signifies the moment of the mean conjunction
of this month. Consequently the time inverval from line to line
represents always the same amount of one mean synodic month.

The arithmetical structure of the second column is simple to
analyze. All the numbers of the first three lines end in 57,58.
Then follow six lines ending in 21,22 and finally we have four
lines which have 36,38 as terminal figures. Thus we need only
concentrate our attention on the first two places. The first three
lines show a fixed decrease of 18 in the second place:

$$28,55 \qquad 28,37 \qquad 28,19$$

these influences and to develop a theo
diction of their combined effects. Eppin
berger have indeed demonstrated that the
the Seleucid period follow in all essential ste
analysis.

Before turning to the description of these ep
observe that the solution of the problem of first

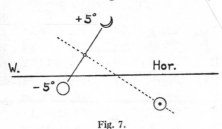

Fig. 7.

permits the solution of some other problems which were a
of great interest. First of all, the day by day positions of su
and moon can easily be established as soon as the laws which
determine the variation of solar and lunar velocity are known.
Thus it is not surprising to find tables which give the daily motion
of sun or moon. Secondly, one can solve the problem of last
visibility of the moon by applying essentially the same argument
to the eastern horizon and the rising of sun and moon. Finally,
both the first and last visibility require as a preliminary step the
knowledge of the moments of conjunction which fall in the middle
of the interval of invisibility. Exactly the same considerations
lead to the computation of the moments of opposition. If we
combine this knowledge with the rules which determine the
latitude of the moon, we can answer the question when the
moon will be close to the ecliptic at oppositions or conjunctions.
In the first case we can expect a lunar eclipse, in the second a
solar eclipse. Thus it is only a logical step which leads from the
computation of the new moons to eclipse tables which we find
derived from the ephemerides.

I hope that this superficial summary of the main results and
problems of the lunar theory will suffice to give an impression
of the inner consistency and the truly mathematical character of
this theory. The following discussion of some details is not only

The next group shows an increase of 18 from line to line:

28,19   28,37   28,55   29,13   29,31   29,49

Then follows again a decreasing sequence:

29,56   29,38   29,20   29,2

with difference of 18. If we plot these numbers in a graph with equidistant points representing the consecutive lines, then we

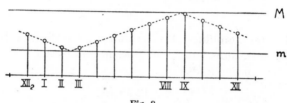

Fig. 8.

obtain a sequence of points which lie on straight lines of alternating slope $\pm$ 18 (cf. Fig. 8). We call such sequences "linear zigzag functions".

The straight lines intersect in a maximum value M and a minimum value m which can easily be computed from our table. One finds

$$M = 30,1,59,0$$
$$m = 28,10,39,40$$

From similar tables one can demonstrate that the same extremal values were used. Consequently our linear zigzag function is bounded by a fixed maximum M and a fixed minimum m and therefore forms a periodic function of amplitude

$$\Delta = M - m = 1,51,19,20$$

and mean value

$$\mu = \tfrac{1}{2}(M + m) = 29,6,19,20.$$

Finally we introduce the concept of "period" P. The abscissa in our graph is divided into equidistant steps, each of which represents a mean synodic month. We now can ask for the distance between two consecutive points of maximum (or minimum)

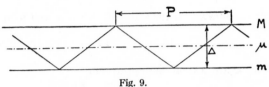

Fig. 9.

measured in these units which represent mean synodic months (cf. Fig. 9). By a simple geometrical argument one finds that

$$P = \frac{2\,\Delta}{d}$$

where Δ is the amplitude $M - m$ and d the difference from one line to the next. Substituting our present numbers $\Delta = 1,51,19,20$ and $d = 18,0,0$ one finds

$$P = \frac{3,42,38,40}{18,0,0}.$$

We can cancel common factors in numerator and denominator or we can express this ration as a sexagesimal fraction. Thus we obtain finally:

$$P = \frac{2,46,59}{13,30} = 12;22,8,53,20.$$

In other words we have shown that two consecutive maxima or minima in the zigzag function of the second column are separated by 12;22,8,53,20 mean synodic months or, roughly, by slightly more than $12\frac{1}{3}$ months.

The astronomical significance of the second column is revealed by means of the third column. The first line of the third column is obviously to be interpreted as a point in the ecliptic of longitude ♈ 22;8,18,16. If we add to this the value 28;37,57,58 which we find in the second line of the second column we obtain

♈ 50;46,16,14 = ♉ 20;46,16,14

and this is the longitude found in the second line of the third column. The same rule applies for all subsequent lines and we can therefore say that the second column contains the differences of the third column. Combining this result with the fact that the first column contains the dates of the consecutive conjunctions, we can say that the third column gives the monthly longitudes of the moon and also of the sun because we are dealing with conjunctions. The second column gives the monthly progress of the sun or the solar velocity. Thus we have reached the important result that the ephemeris under discussion represents the yearly variation of the solar velocity by means of a linear zigzag function.

Another important item of information is contained in the value we have found for the period P of this zigzag function, namely

$$P = \frac{2,46,59}{13,30} = 12;22,8,53,20 \text{ months.}$$

This shows not only the value which was adopted here for the length of the year, measured in mean synodic months, but we can read this relation also in the form

$$13,30 \text{ years} = 2,46,59 \text{ months}$$

or

$$810 \text{ years} = 10019 \text{ months.}$$

It seems as if this relation would imply the use of observational records going back more than 800 years. This conclusion is, however, too hastily drawn. First of all, it can be shown that other columns of the same type of ephemerides are based on the simpler relation

$$P = \frac{46,23}{3,45} = 12;22,8 \text{ months}$$

or

$$225 \text{ years} = 2783 \text{ months.}$$

But neither can this relation be taken as the direct result of observations. The period of a zigzag function is given by the quotient of $2\varDelta$ and d where $\varDelta$ is the amplitude $M - m$ and d the difference. The values of d in a linear zigzag function are usually simple numbers—in our example 18,0,0 and not, perhaps, 17,59,59—as is easy to understand in view of the practice of computing an ephemeris, where the value of d has to be added or subtracted in every single line. The accuracy of the value of $\varDelta = M - m$ is reflected in the number of sexagesimal places of M and m and thus of all intermediate places. Again it is reasonable to choose conveniently small numbers for M and m and $\varDelta$. In other words the value of P depends on small correction in the values of $\varDelta$ and d and does not depend solely on the initial empirical relation between the number of years and the corresponding number of months.

This situation is typical throughout Babylonian astronomy. The ephemerides alone are never a reliable source for the investigation of the basic empirical facts. At present it is completely impossible to write a "history" of Babylonian astronomy in its latest phase. All we do have is the ephemerides in a form excellently adapted to practical computation and to predicting new moons, eclipses, etc. We do not know, however, which empirical elements were actually used for the determination of the basic parameters nor are we able to retrace the steps by means of which the theory was formed.

49. The example of an ephemeris which we have quoted in the preceding section is based on the assumption that the variation of the solar velocity follows the scheme of a linear zigzag function. All texts which show this pattern will be called texts of "System B". In contrast to this we classify ephemerides as belonging to a "System A" if the solar velocity is assumed to be constant on two complementary arcs of the ecliptic in the following way. From ♍ 13 to ♓ 27 the sun moves 30° in each mean synodic month; from ♓ 27 to ♍ 13 with a motion of 28;7,30° per month

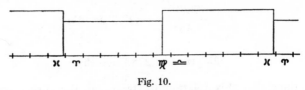

Fig. 10.

(cf. Fig. 10). It is easy to show that this corresponds exactly to the relation which we mentioned at the end of the last section, namely,

$$1 \text{ year} = 12;22,8 \text{ months}$$

and which also occurs otherwise in ephemerides of both systems.

Obviously it seems to be a much more natural assumption to let the solar velocity vary continuously instead of having discontinuous variations at ♍ 13 and ♓ 27. Nevertheless Kugler has seen that System A is in general more primitive and therefore older than System B, and this has been confirmed by subsequent study. In particular it can be shown that the assumption of a solar movement of the type of System B leads to rather complicated consequences and it is on this basis that System A adopted its cruder pattern. Here again we meet with the fact that purely mathematical considerations exercise an essential influence on the details of the theory behind which the original empirical data and general concepts are veiled from our sight.

Though the chronological priority of System A seems to be well established we have no means to determine the date of origin of either one. It is furthermore a curious fact that both systems were simultaneously used during the whole period (from about 250 B.C. to about 50 B.C.) for which ephemerides are preserved. This coexistence of two different methods of computing ephemerides is not a matter of "schools" in so far as both "systems" are attested both at Babylon and at Uruk, the places of origin of the only two archives to which we can safely assign our texts. It is difficult to explain why both methods were kept alive in spite of the fact that System B was certainly an improvement over System A in several respects. In the planetary theory a still higher multiplicity of procedures exists simultaneously, very much contrary to our modern scientific habits.

50. We now turn to a rapid summary of the lunar ephemerides without explicitly attempting to derive our statements from the textual material or to analyze in detail the general theory which formed the basis upon which the numerical procedures were built. Our sketch includes both systems, leaving aside variants which occur especially in System B.

The general arrangement of all ephemerides is identical. Each line represents a month, each column a specific "function" like solar velocity, lunar velocity, etc. We denote these functions by capital letters, which we also use for the corresponding columns. The majority of lunar ephemerides cover one year but we also have texts which concern two or even three years. The general appearance of such a text with its different columns is shown on Pl. 7b. The columns of ephemerides always proceed from left to right.

The first column in all ephemerides is the column of dates T, giving the year of the Seleucid era and the consecutive months (cf. the example on p. 110). Because the edge of a tablet is particularly exposed to destruction, one often meets the problem of restoring the date of an ephemeris. This can be done by continuing preserved columns until one reaches the corresponding column of another dated text. In this very way it is possible to show that all texts of System A form one consistent set of ephemerides throughout the whole interval (of two centuries) at our disposal. System B, however, shows a much lower degree of uniformity.

The next column, Φ, is peculiar to System A only. Its period is identical with the period of the variable lunar velocity and its units are time degrees. It is used for the computation of the variable length of the synodic month (column G) under the preliminary assumption of constant solar velocity. The details of the construction of this linear zigzag function Φ are not yet clear but it is now certain that it is related to the 18-year cycle, the socalled "Saros" of 223 mean synodic months. This period is slightly shorter than 239 anomalistic months and therefore also the length G of the synodic month almost repeats itself after one Saros. The slight difference of time in the length of two months one Saros apart is the difference of Φ. From this change of G after one Saros can then be found the corresponding change of G from month to month. This is an interesting case of an important method of ancient astronomy: the accumulated error after the lapse of a relatively short approximate period (here 18 years or 223 months) is used to determine the correction from step to step (here a single synodic month).

The next column is column A of System B and gives the solar velocity as described in our example of p. 110, column II. From this is derived column B, containing the longitudes of the moon and of the sun at conjunction or, for full moons, the longitudes of the moon, the sun being 180° distant. In System A, column B is derived without explicit mention of the velocity (column A) because in System A only two velocity values are used, and thus there was no reason to repeat them in a special column.

The subsequent columns, C and D and variants, give the length of daylight or night corresponding to the solar longitude

of column B. The functions C and D are computed according to
independent arithmetical schemes designed to represent quan-
titatively the variation of the length of daylight during the year.
The underlying problem is one of spherical trigonometry but it
was solved here by arithmetical devices similar to the approximations of a sinusoidal curve by a linear zigzag function.

The two following columns, E and Ψ, describe the variations
of the latitude of the moon and the magnitude of eclipses. As
we have remarked previously, the consecutive lines of an ephemeris refer to the consecutive conjunctions or oppositions. If the
latitude of the moon is known for these moments, one is able
to judge the possibility of an eclipse and to compute, if necessary,
its magnitude. The latitude itself is again found by means of
zigzag functions. The "eclipse magnitude" is expressed in a
slightly different way than is customary today, but it is easy to
transfer it directly into a measure for the depth of immersion of
the lunar disc into the shadow. It is interesting to see that this
quantity was computed in many ephemerides for every month
and not only for every sixth (or perhaps fifth) month when an
eclipse is possible. In other words, a method had been developed
for computing "eclipse magnitudes" as a function of the latitude
such that the numbers obtained gave the size of the eclipse
correctly for real eclipses. For non-ecliptic conjunctions, however, these values behave exactly as if the distance from the
shadow was introduced as eclipse magnitude, allowance being
made for negative distances if the shadow is not reached, positive
values giving the depth of immersion for a real eclipse. This
shows a remarkably abstract attitude in the Babylonian procedure, which unhesitatingly introduces quantities for purely
mathematical convenience, in principle very much the same as
the use of complex numbers in modern mechanics.

The next column, F, gives the variations of the lunar velocity
in a form similar to column A for the solar velocity. In column G
we find the length of the synodic months under the assumption
of a constant solar velocity but a variable lunar velocity as
indicated by column F. At the beginning of our discussion we
had to make the assumption that consecutive lines represented
mean conjunctions, separated by the mean length of a synodic
month. This mean length would be produced by the conjunctions

of a sun and a moon, each moving with its own mean velocity. In column G this assumption is partially abolished insofar as only the sun is moving with its mean velocity and the answer is given to the question how much a given variation in the lunar velocity influences the spacing between consecutive conjunctions. Obviously G will show the same period as F; the value of G will be small and the month will be short if the moon moves fast, i. e., near the maximum of F. This is indeed the relation between F and G.

The next step, J, gives the necessary corrections to G because of the variable solar velocity. In System B, column J is a difference sequence of second order due to the fact that column A is a linear zigzag function. Here it becomes evident why the inventor of System A preferred to assume a simple step-function for the solar velocity; the corrections for variable solar velocity are much more complicated in System B than in System A. After the correction J has been found, the algebraic sum K of G and J gives the length of the synodic month as it results from the variability of both sun and moon. If the moment of one conjunction is known, one need only add to it the amount of K found for the length of the following synodic month and one obtains the moment for the conjunction of the next month. Actually a slight complication is introduced here by the use of the Babylonian calendar, which requires that the beginning of a day be counted from actual sunset and not from midnight. Hence a correction for the transformation from midnight epoch to evening epoch is required. This can be done easily by means of columns C and D which give us the length of daylight or night. After this transformation is carried out, we obtain in column M the dates and moments of all consecutive conjunctions referred to sunset. Thus the first goal of the lunar theory has been reached: the moments of the actual conjunctions or oppositions are known.

51. For the computation of eclipses no more information is needed than has been collected thus far. We have in column M the time of the conjunction or opposition expressed in its relation to sunset or sunrise. From column Ψ, we know the distance of the moon from the shadow. The ephemerides and eclipse tables show with full clarity that one knew that solar and lunar eclipses were subject to the same conditions, namely, sufficiently small

latitude near new or full moon. The problem of determining these moments and of describing the motion in latitude was solved very successfully by means of arithmetical methods. Consequently one obtained quite satisfactory results for the prediction of lunar eclipses (cf. Fig. 11[1])). For a solar eclipse we should know more; specifically, we should be able to judge whether the vertex of the shadow cone touches our particular locality, assuming all other circumstances are favorable for an eclipse. This problem can be solved only if sufficiently accurate information about the actual distances of sun and moon from the earth are available, together with a correct knowledge of the relative sizes of these bodies. There is not the slightest reference to any of these quantities in Babylonian texts. Tables for solar eclipses are computed exactly like the tables for lunar eclipses with no additional columns corresponding to "parallax", i. e., quantities depending on the above-mentioned distances and sizes. Consequently the Babylonian texts do not suffice to say anything more than that a solar eclipse is excluded or that a solar eclipse is possible. But they cannot answer even approximately the question whether a possible solar eclipse will actually be visible or not. One has to remember that this is the state of affairs during the last period of Mesopotamian astronomy, from about 300 B.C. to 0. Before 300 B.C. the chances for the correct prediction of a solar eclipse are still smaller. At all periods, exclusion of an eclipse of the sun is the only safe prediction that was possible.

52. The remaining part of the ephemerides concerns the fundamental problem of the lunar calendar: to determine the evening of first visibility after conjunction when the new crescent again becomes visible. We have already discussed (p. 107) the three major factors which determine the visibility of the new crescent at a sunset following conjunction, namely, elongation, variable inclination between ecliptic and horizon, and latitude of the moon. It is exactly these three quantities which are found in the columns O, Q, and R of ephemerides of System B. Column O for the elongation is preceded by a column N which gives the

---

[1]) Fig. 11 illustrates the results obtained for the magnitude of lunar eclipses, expressed in digits such that 12 means totality. The omission of modern values indicates that no eclipse would have been visible, according to modern computation. The same is indicated by ancient values ≤ 0.

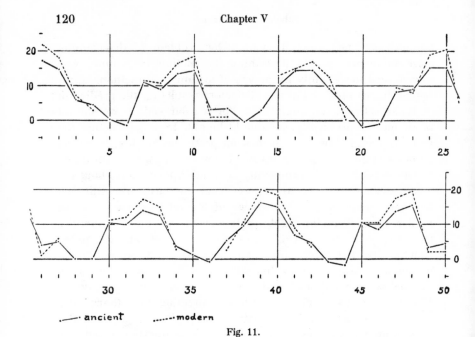

Fig. 11.

time difference between the moment of conjunction and the subsequent sunset at which the first crescent might be expected. For
this particular evening one computes how long the new crescent
will be above the horizon after sunset. If the resulting time
difference between sunset and the setting of the moon is long
enough to secure visibility, then the initial guess was right and
the evening which starts the new month is known. If the resulting
value seems too high, the computation has to be repeated for
one day earlier. If the first result seems too low, a new value
must be found for 24 hours later. In some cases alternative results
are recorded in the final column P, corresponding to either a
29-day month or a 30-day month.

Though the general outlines of this part of the lunar theory
are clear, many details are still obscure, chiefly because of the
difficult terminology of the procedure texts and the rounding-off
of the numbers involved in the actual ephemerides. An added
difficulty results from the fact that the ephemerides of System A
do not give any of the columns N, O, Q, and R but list only
the final result P. Nevertheless, a few additional facts can be

established. First of all, it is clear that the determination of the influence of the variable angle between ecliptic and horizon is a problem of spherical trigonometry, and the same holds for the influence of the lunar latitude. Exactly as in the case of the length of daylight, this problem was solved by means of fixed arithmetical schemes. The procedure texts give lists of coefficients by which the elongation has to be multiplied in order to obtain for different solar longitudes the proper amount of difference in time for setting, and a similar device is followed for the latitude. The main difficulty for us consists in discovering on what grounds the decision was made as to whether a given value in the final column P was sufficient for visibility or not. It seems as if not P alone had been used, but the sum of the elongation O and the value of P. Indeed the brightness of the new crescent depends essentially on the width of the illuminated sickle of the moon, and this width is proportional to the elongation. Thus it is reasonable to say that even a small value of P, caused by closeness of the moon to the horizon, might be compensated for by a greater brightness of the sickle, and, vice versa, a very small sickle might not be visible even at a relatively great distance from the horizon. Thus the sum of O and P is indeed a very reasonable parameter to be used as a criterion of visibility.

53. Concluding our summary of the lunar theory, we must still mention the texts which concern the daily motion of the sun and the moon. Indeed, there exist ephemerides which give the longitude of the sun from day to day, assuming a constant mean velocity of $0;59,9°$ which is slightly too high a value.

Similar ephemerides also exist for the moon, though under the assumption of a variable lunar velocity. This variation is, as usual, expressed in the form of a linear zigzag function. The mean velocity is assumed to be $13;10,35°$ per day, a value which appears again and again in ancient and medieval astronomy. The extremal values are $m = 11;6,35°$ and $M = 15;14,35°$ from which one derives a period

$$P = \frac{4,8}{9} = 27;33,20 \text{ days.}$$

This indicates that one "anomalistic" month is given the length of $27;33,20$ days or, expressed as a relation between integers, that 9 anomalistic months contain 248 days. This relation is not quite accurate, as can be shown by comparison with the previously discussed ephemerides. In the latter we find that column F for the lunar velocity is based on the relation

$$4,29 \text{ anomalistic months} = 4,11 \text{ synodic months.}$$

Substituting in this equation the value 27;33,20 days for the length of the anomalistic month we obtain for the synodic month a value close to 29;31,54 days. But from column G one derives for the mean synodic month the length of 29;31,50,8,20 days, which is again one of the classical parameters of ancient and medieval lunar theory. Hence it is clear that 27;33,20 is slightly too high a value, caused by the desire to obtain conveniently short numbers for the parameters of the zigzag function for the daily motion. We shall come back to this remark in our last chapter (p. 162).

54. Before describing the Babylonian planetary theory, we shall discuss the main features of the apparent movement of the planets from a modern point of view. We know that the planets move on ellipses around the sun, the earth being one of them. We shall derive from these facts the apparent motions as seen from the earth. In order to simplify our discussion, we shall replace all orbits by circles whose common center is the sun. The eccentricities of the elliptic orbits are so small that a scale drawing that would fit this page would not show the difference between the elliptic and the circular orbits.

We utilize furthermore the fact that the dimensions of our planetary system are so minute in comparison with the distances to the fixed stars which constitute the background of the celestial sphere that we commit no observable error at all if we keep either the sun or the earth in a fixed position with respect to the surrounding universe. Hence we will proceed in the following way. We shall start with the circular motion of the planets around the sun and then keep the earth fixed and ask for the resulting motion with respect to the earth. This will answer our question concerning the planetary phenomena.

The first step is absolutely trivial. We know that the earth is a satellite of the sun, moving around it once in a year. In order to obtain the appearances seen from the earth we subtract from

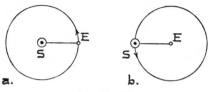

Fig. 12 a–b.

all motions the motion of the earth. Thus we see that by arresting the motion of the earth we obtain the appearance that the sun moves around the earth once per year. Its apparent path is called the ecliptic (cf. Fig. 12 a and b).

Secondly we consider an "inner" planet, Mercury or Venus, which moves closer to the sun than the earth (Fig. 13 a). If we stop the earth we need only repeat Fig. 12 in order to obtain again the motion of the sun. The orbit of the planet remains a circle with the sun in its center. Hence the geocentric description of the motion of an inner planet is given by a planet which moves on a little circle whose center is carried on a larger circle whose center is the earth. The little circle is called an "epicycle", the large circle is the "deferent".

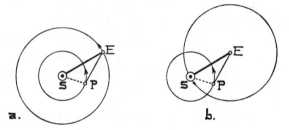

Fig. 13 a–b.

Finally we have an "outer" planet, Mars, Jupiter, or Saturn, whose orbit encloses the orbit of the earth (Fig. 14 a). From the earth E the planet P appears to be moving on a circle whose center S moves around E. Thus we have again an epicyclic motion (Fig. 14 b). In order to establish a closer similarity with the case of the inner planets we introduce a point C such that the four points S, E, P, and C always form a parallelogram. SP is the radius of the planetary orbit; because $EC = SP$ we

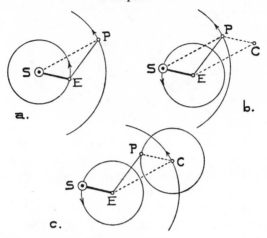

Fig. 14 a–c.

see that C lies on a circle with center E. Similarly ES is the radius of the solar orbit, and, because $ES = CP$, we see that P lies on a circle around C. Thus the planet P moves on an epicycle whose center C travels on a deferent whose center is E (Fig. 14 c). Thus we have established an exact analogue to the case of the inner planets. In both cases the planet has an epicyclic movement. In the case of the inner planets the center of the epicycle coincides with the sun. For the outer planets the center C of the epicycle moves around E with the same angular velocity as the planet moves around the sun, while the planet P moves on the epicycle around C with the same angular velocity as the sun moves around the earth.

In order to avoid misunderstandings, I shall repeat once more the assumptions upon which our above results rest. These assumptions were (a) that the planetary orbits are circles with the sun in their common center; (b) that all planetary orbits lie in the same plane. Accepting these two assumptions we have seen that the planetary orbits with respect to the earth consist of epicycles whose centers move with uniform velocity on deferents having the earth as center. In other words, if we disregard the small eccentricities of the planetary orbits, and if we also neglect the small inclinations of these orbits, then the epicyclic motion gives a correct description of the planetary orbits with respect to the earth. Indeed it is only a matter of mathematical convenience whether one computes first the longitudes of the earth and the planets heliocentrically and then transforms to geocentric co-ordinates, or whether one carries out this transformation first and then operates with epicycles.

For a finer theory of the planetary phenomena the above assumptions are too crude. It is easy, however, to see in what directions one should move in order to reach higher accuracy. The eccentricity of the orbits can be taken into consideration by assuming slightly eccentric positions of the earth with respect to the centers of the deferents. The latitude can be accounted for by giving the epicycles the proper inclination. Both devices were followed by the Greek astronomers.

55. We have now seen that the planets move with respect to the earth on epicycles. This makes it particularly simple to understand the main features of the planetary motions as seen from the earth.We begin again with an inner planet. Its angular velocity about the center S of its epicycle (cf. Fig. 15) is greater than the angular velocity of S about the earth E. If the planet P is on the part of its epicycle which is removed from the earth, the motion of P is added to the motion of S and the planetary

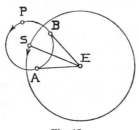

Fig. 15.

motion appears greater than the motion of S. We call this the "direct" motion. Between A and B, however, the planet moves backward faster than its epicycle is carried forward[1]), thus it appears to be "retrograde".

The same figure allows us also to describe the visibility conditions. If the planet P and the sun S are seen in the same, or

[1]) It is easy to see that the points A and B lie somewhat inside the two points where the lines from E are tangential to the epicycle.

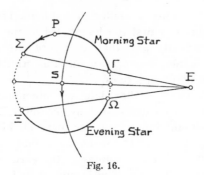

Fig. 16.

in nearly the same, direction from E, the planet is invisible
because of the brightness of the sun. Thus a certain "elongation"
of the planet from the sun is required to make the planet visible.
Fig. 16 shows that the arc of invisibility between Σ and Ξ near
"superior conjunction" is much greater than between Ω and Γ
(near "inferior" conjunction). The visible arc from Γ to Σ rises
before the sun; thus the planet is "morning star". The arc from
Ξ to Ω sets after the sun; thus the planet is "evening star". Fig. 17
describes the same phenomena once more in a graph with the
abscissa representing time whereas the ordinates represent geocentric longitudes. The straight line represents the motion of
the sun.

In similar fashion one obtains for an outer planet a graph as
given in Fig. 18. Now the motion of the planet is slower than
the motion of the sun. Retrogradation occurs near opposition, Θ,
when the sun and planet are seen in opposite directions from the
earth. Consequently the retrogradation of an outer planet is fully
visible in contrast to that of an inner planet, where a part of the
retrograde motion becomes invisible near inferior conjunction.
An outer planet becomes invisible only once in each cycle: near
conjunction, Ω to Γ. The points Φ and Ψ, where direct motion
changes to retrograde motion and vice versa, are called the "first"
and "second" stationary points respectively.

56. It is in the theory of the planets that the contrast between
the Babylonian approach and Ptolemy's theory as presented in
the Almagest becomes most visible. In the Ptolemaic theory a
definite kinematic model is assumed, based on epicyclic motion,
which closely corresponds to the description of the planetary

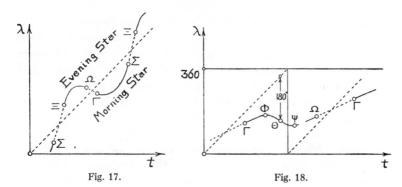

Fig. 17.   Fig. 18.

motion given in the preceding sections. Thus the geocentric longitude of the planet can be computed for any given moment t. It is then a secondary problem to determine those values of t for which the planet is in one of the characteristic phenomena which we denoted by Greek letters.

The Babylonian method follows the exactly opposite arrangement. The first goal consists in determining the "Greek-letter phenomena", and thereafter the longitude of the planet for an arbitrary moment t is found by interpolation.

This difference in approach is, of course, the result of the historical development. The Babylonians were primarily interested in the appearance and disappearance of the planets in analogy to the first and last visibility of fixed stars—e. g. Sirius—and of the moon. It was the periodic recurrence of these phenomena and their fluctuations which they primarily attempted to determine. When Ptolemy developed his planetary theory, he had already at his disposal the geometrical methods by means of which the solar and lunar anomalies were explained very satisfactorily, and similar models had been used also for an at least qualitative explanation of the apparent planetary orbits. Thus it had become an obvious goal of theoretical astronomy to offer a strictly geometrical theory of the planetary motions as a whole and the characteristic phenomena lost much of their specific interest, especially after the Greek astronomers had developed enough observational experience to realize that horizon phenomena were the worst possible choice to provide the necessary empirical data.

57. Whatever phenomenon the Babylonian astronomers wanted to predict, it had to be determined within the existing lunar calendar. Suppose one had found that a planet would reappear 100 days from a given date. What date should be assigned to this moment? Obviously one should know whether the three intermediate lunar months were, perhaps, all only 29 days long, or all three were 30 days long, etc. This question could be answered perfectly well by lunar ephemerides whose goal it was to determine whether a given month was 29 or 30 days long. But planetary phenomena proceed very slowly. One single table for Jupiter or Saturn could easily cover 60 years and more. To determine calendar dates so far in advance would have meant the computation of complete lunar ephemerides for several decades. Furthermore, the actual computation of the planetary motion had to be based, in any case, on a uniform time scale. All these difficulties were at once overcome by a very clever device. One used as unit of time the mean synodic month and divided it in 30 equal parts. The Babylonians seem not to have had a special name for these units, referring to them simply as "days". Modern scholars have used the term "lunar days"; I shall use the corresponding term of Hindu astronomy, namely, "tithi".

The fact that the Babylonian calendar was a strictly lunar calendar has the effect that the total duration of a number of calendar months will not deviate more and more from the corresponding total of mean synodic months. Dates expressed in tithis will never be far off from real calendar days, usually not more than ±1 day. Thus the Babylonian astronomers in their computations simply identified the results given in tithis with the dates in the real calendar. This is the standard procedure for all planetary texts.

The use of tithis implies that one did not try to reach, for the planetary phenomena, the same accuracy which was obtained in the lunar theory. While one went to great lengths to determine all possible influences upon the first and last visibility of the moon, we find no similar devices used for the planetary phenomena. The latitude of the planets, for instance, is nowhere taken into consideration for the planetary ephemerides. On the other hand, several concurrent "systems" are used simultaneously, as we have seen also in the case of the two systems of the lunar theory. The different systems of the planetary theory are obviously modeled after the two main systems of the lunar theory. They either operate with step functions (type "A") or with linear zigzag functions (type "B"). The variations within ephemerides of type A consist in the use of different numbers of steps for each period. There exists, for instance, a theory of Jupiter with only two ecliptical zones of different velocity, while another method operates with four zones, two intermediate steps being inserted between the extremal values of the previous model. The variations of texts of type B consist in small changes (rounding-off) of parameters; similar variations are also known in System B of the lunar theory.

The basic idea of all planetary ephemerides is, however, the same. It consists in the separate treatment of each characteristic phenomenon by itself as if this phenomenon were an independent body moving in the ecliptic.

Let us consider, as an example, the first appearance Γ of Mercury as a morning star. We assume that we are given (by observation or by previous computation) the moment t_0 and the longitude λ_0 of Mercury when it again became visible in the morning after a period of invisibility at inferior conjunction (cf. Fig. 16 and 17 p. 126 f.). We call this point Γ_0 in our diagram Fig. 19. If both the sun and Mercury would move with constant velocity and if this motion would fall in the equator, then subsequent morning appearances Γ_1, Γ_2, ... would be spaced equidistant in the diagram, keeping a fixed distance from the graph of the solar motion. Actually, however, these assumptions are not satisfied. Therefore the spacing of the points Γ_0, Γ_1, Γ_2, ... shows periodic irregularities. The Babylonian theory tries to describe these irregularities precisely in the same fashion as the solar and lunar theory described the variable velocity of these bodies. Hence for an ephemeris of type A the ecliptic is divided into zones such that the progress of the phenomenon Γ in each zone is given by the same amount, with discontinuous changes at the boundary.

For Mercury and Γ we have three zones with discontinuities at ♌ 1, ♑ 16, and ♊ 0. Suppose that Γ_0 is given to be in ♌ 17. The velocity in the zone which stretches from ♌ 1 to ♑ 16 is

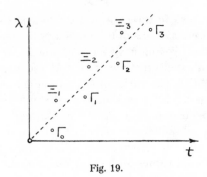

Fig. 19.

$1,46°$. Consequently Γ_1 will be in ♌ $17 + 1,46 = $ ♌ $2,3 = $ ♐ 3. This point still lies within the same zone; thus we again add $1,46°$ and obtain ♐ $1,49 = $ ♓ 19. In this step we have crossed, however, the boundary ♑ 16 and have entered the second zone by an amount of $1,3°$. In the second zone the velocity is no longer $1,46°$ but $2,21;20°$ or $\frac{1}{3}$ greater than the previous velocity. Thus we have to raise the arc of $1,3°$ also by $\frac{1}{3}$ of its amount, i. e., by $21°$. Thus Γ_2 will not be ♓ 19 but ♓ $19 + 21 = $ ♈ 10.

In the same fashion, subsequent positions providing us with all longitudes of Γ_1, Γ_2, ... can be found. The corresponding dates, expressed in tithis, are determined by a simple rule which makes the time differences linearly dependent upon the differences in longitude. Thus two columns of an ephemeris can be computed, one giving the longitudes, the other the dates for consecutive Γ's.

How one determined the parameters which characterize the distribution of the Γ's is a difficult question which cannot be answered completely. Only this much is easy to see: some counting of the number of Γ's must have been made such that an integer number of first appearances of Mercury as morning star corresponds to an integer number of years. The above mentioned parameters are based on the relation that 848 years contain 2673 risings of Mercury. Exactly as in the lunar theory, no historical conclusions can be based on these numbers. The size of the zones, the particular zonal velocities and their ratios must be comparatively handy numbers, and the period relation derived from these numbers reflects nothing more than the final compromise between empirical facts and computational requirements.

The next step in the computation of the phenomena of Mercury consists in finding longitudes and dates for all consecutive first appearances as evening star beginning with a given point $\varXi_0$. The principle of the procedure is perfectly analogous to the procedure for the $\varGamma$'s, but the zones and velocities are different. The discontinuities are now located at $\simeq$ 26, $\times$ 10, and ϖ 6 and the velocities are 1,46;40°, 1,36°, and 2,40°, as compared with 1,46°, 2,21;20°, and 1,34;13,20° in the previous case. The period relation is now expressed by the equivalence of 480 years and 1513 risings of Mercury. This drastically illustrates our previous remark that no historical conclusions can be drawn from these relations because it is obviously absurd that the observations of the $\varGamma$'s should extend for centuries farther back than the observations of the $\varXi$'s. It is furthermore clear that the two periods should be identical because every $\varGamma$ must always be followed by exactly one $\varXi$, and vice versa. This fact was of course evident to the Babylonian astronomers and the two periods deviate from each other only as 3;9,7,38,.. from 3;9,7,30. It is only the adjustment of the determining parameters of the zones which causes an apparent discrepancy, the effect of which was negligible in practice.

The preceding steps provide us with all $\varGamma$'s and $\varXi$'s. We still have to find the $\varSigma$'s and $\varOmega$'s, that is, the corresponding settings of Mercury. One might expect to find two additional schemes which yield these data in the same fashion as the $\varGamma$'s and $\varXi$'s were found before. This is, however, not the case. Kugler suspected from the fragmentary ephemerides at his disposal that the $\varSigma$'s were computed from the $\varGamma$'s by means of fixed additive amounts depending only on the longitude of $\varGamma$; and, similarly, the $\varOmega$'s from the $\varXi$'s. Kugler's hypothesis has been fully confirmed by texts from Uruk. Tables were computed which give for every degree of every zodiacal sign the amount of longitude and the number of tithis which must be added to a given $\varGamma$ (or $\varXi$) in order to find the subsequent $\varSigma$ (or $\varOmega$). In other words we have fixed arithmetical schemes which determine as function of λ the relationship between consecutive risings and settings of Mercury.

Fig. 20 illustrates the graphical representation of one of these curves[1]). One point needs special emphasis. The region near the

[1]) The longitudes of $\varOmega$ and $\varXi$.

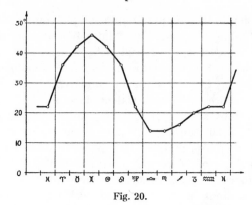

Fig. 20.

minimum indicates that in this part of the zodiac the corresponding disappearance should follow very shortly after the preceding
appearance. The text declares this whole zone as cases of invisibility. In other words, for this region the planet is never
visible at all though one "rising" and one "setting" should be
theoretically counted in order to maintain the correct period
relation. In the representation of our graph (Fig. 17 p. 127) there
exists a zone of the zodiac where one whole bulge remains so
close to the sun that it should be dotted as invisible.

It is interesting to remark that the most up-to-date modern
tables for the computation of risings and settings of Mercury are
based on ephemerides which contain dates and longitudes for
actually invisible phenomena. In the ancient texts these cases are
denoted by an ideogram LU whose significance became clear
only when the computation of the Σ's and Ω's was fully understood. Without this knowledge, however, one had to assume an
extremely high visibility for the horizon in Babylon in order to
cover all recorded cases, not realizing that these records contained invisible and visible risings alike. Consequently our
modern tables assume a much too high degree of visibility of
Mercury in Mesopotamia and therefore yield results which have
only a very general resemblance to the facts and are definitely
wrong in the critical cases for which they were computed.

58. The computation of ephemerides for the other planets
follows in general the ideas which we have illustrated in the case
of Mercury.

The discovery, among Pinches's copies (cf. p. 105), of several planetary procedure texts and ephemerides disclosed the existence of a much greater variety of methods in detail than we had assumed on the basis of the material accessible to Strassmaier and Kugler. The following is a short summary of the main systems according to my present knowledge.

For all planets the central problem consists in establishing the variations in the amount of the synodic arcs between phenomena of the same kind as function of the ecliptic. If the synodic arcs are known, then the synodic times are easily found since they are the time required for the sun to travel between two consecutive phenomena of the same kind. Though some correction had to be applied to this basic principle, we need not discuss the details here. For the longitudes, however, the variety of methods is of real interest since it shows that great efforts were made to describe properly the appearances of the planetary motion.

As is to be expected, the most regular variation in the synodic arcs is found in the case of Saturn. We know of two systems, one proceeding with two zones of constant synodic arc ("System A"), one with synodic arcs which vary according to a linear zigzag function ("System B"). Both systems concern all five phenomena: Γ, Φ, Θ, Ψ, and Ω.

For Jupiter we know of two closely related procedures following System B, for the same five phenomena as in the case of Saturn. We also have, however, a variety of methods of type A: two zones, four zones, three zones, and variants as to location and amount of the discontinuities. Only a fraction of these systems is attested in actual ephemerides, though they are relatively numerous for Jupiter.

While there is, in principle, only little difference between the theory of Jupiter and that of Saturn, we find a different procedure followed in the case of Mars. We know of a six-zone System A (all zones being two zodiacal signs long), but applied only to first and last appearance, Γ and Ω, and to first station, Φ. For the retrograde arc, however, four different methods are known for the determination of the amount of the retrograde arc from Φ to opposition Θ, and probably equally many existed for the arc from Θ to Ψ. In other words, we have here a situation similar to the procedure that we described in the case of Mercury. After the position of Φ has been found by System A, the position of Θ and Ψ are determined by means of pushes whose (negative) values depend on the longitude of Φ. The position of the subsequent phenomenon, Ω, is entirely independent of the size of the preceding retrogradation.

Venus is treated quite differently. Here one utilizes first of all the fact that in eight years Venus completes five synodic periods which, in the mean, are only $2\frac{1}{2}°$ short of the initial position in the zodiac. Now for each of the six characteristic phenomena (Ξ, Ψ, Ω in the evening, Γ, Φ, Σ in the morning), numerical rules are given which indicate the synodic arcs and times in a sequence of five consecutive phenomena of the same kind. Thus we are given a table of 5 times 6 or 30 pushes for the longitudes and equally as many for the times. The total of 5 pushes for longitudes is such that it results in a deficit of $2;30°$, which is what it should be for each cycle. Similarly, the dates recede by $4;10$ tithis after the 99 lunar months which correspond to 5 synodic periods. This general procedure is known in at least two variants, but no process of type A or B is attested so far. Much of the detail escapes us since our sources concerning Venus are particularly fragmentary.

For Mercury we have described in detail a procedure following System A for the appearances Γ and Ξ while the disappearances Σ and Ω were found by means of pushes depending on the longitudes of the preceding appearances. In other words, the lengths of the arcs of visibility are prescribed as function of the longitude of the first appearances. We also know of the existence of a complementary system: the disappearances Σ and Ω are computed first according to a System A (though using zones which are different from the previous ones) and the arcs of invisibility are prescribed as pushes $\Sigma \rightarrow \Xi$ and $\Omega \rightarrow \Gamma$ respectively. In a general way, the results are equivalent to the previous ones, but are different in detail.

For all planets we now know of a variety of methods for computing from one given phenomenon all subsequent phenomena of the same kind. The question arises as to how the initial values were chosen. One could, in principle, assume that one set of phenomena within one synodic period was determined by observation, e. g., longitudes and dates of Γ, Φ, Θ, Ψ, Ω for Saturn or Jupiter, Γ, Φ, Ω for Mars, etc. We know that the Babylonian astronomers were keenly aware of this problem and tried to develop rules which allow us not only to compute consecutive phenomena of the same kind, but also to pass from one phenomenon to its neighbor of another kind. This problem is by no means trivial since in an ephemeris of the type

$$\begin{array}{ccccc}
\Gamma_0 & \Phi_0 & \Theta_0 & \Psi_0 & \Omega_0 \\
\Gamma_1 & \Phi_1 & \Theta_1 & \Psi_1 & \Omega_1 \\
\Gamma_2 & \Phi_2 & \Theta_2 & \Psi_2 & \Omega_2 \\
\end{array}$$

. . . .

all rows are determined as soon as the first row is given. The rules in each later line are thus a consequence of the rule which determines the relations in the first line. In other words, one must check the consistency of a rule which leads from a Γ to Φ to ... Ω with the previously given methods of computing vertically. This problem has been correctly solved in some cases, in other cases the rules are clearly intended only to serve as approximations. In all cases one sees, however, the tendency to restrict the empirical data to a minimum. Indeed for a consistent system of rules for computing in lines as well as in columns one single value suffices for computing all following phenomena. This is the ideal of a mathematical astronomy of the purest kind.

For all planets there arose the final problem of describing their daily motion. We have now in principle reached the knowledge of all longitudes and dates for the typical phenomena. Using our graphs we can say that we know the positions of all points which we have denoted by Greek letters. The problem remains to determine the intermediate curves. Though only few texts are preserved which permit an investigation of this problem, we at least know that interpolation schemes were devised such that one could start from one given value and reach in a number of steps, given by the difference in date, the next characteristic value. These interpolation schemes are built upon difference sequences of second or even third order. Using modern terminology, one may say that one determined simple polynomials which satisfy with sufficient accuracy the conditions which are expressed by the relative position of consecutive characteristic points in our graph of the planetary motion. One can only admire the elegance and skill which is reflected in all these arithmetical

methods. We are still far from a full appreciation of them since
we know so little about the underlying empirical material which
was so skilfully applied to provide the basic parameters of a
real mathematical theory.

59. It is natural to ask who were the astronomers who developed and used this theory. I see no way of answering such
a question satisfactorily. One can do no more than enumerate
the few facts that we know. The texts from which all our information comes were parts of two archives, one in Uruk, one in
Babylon. There is no proof against the existence of other archives
and we are unable to judge the relations between two or more
centers of astronomical activity. We know very little about the
Babylon archive, because the Babylon texts rarely have colophons.

Thus we are almost completely dependent upon the colophons
of the Uruk texts. These colophons follow more or less the following pattern: "Tablet of A, son of B, son of C, descendant of
M; hand of (= written by) R, son of S, son of T, descendant of N.
Uruk, month m, day d, year y (of the Seleucid era), X being
king". Many tablets contain an invocation at the beginning:
"According to the command of the deities Anu and Antu, may
it go well". Some colophons add a curse against whoever removes
the tablet which was written by the scribe, it is said, "for the
prolongation of his days and for the well-being of his posterity",
and occasionally we read that "the informed may show the
tablet to the informed but not to the uninformed".

The investigation of the kinships mentioned in the colophons
allows us to establish two scribal families which were engaged
in writing ephemerides or were their owners (if this is the meaning
of the phrase "tablet of"). One family mentions Ekur-zākir as
their "ancestor", a man who is given the title of "mashmash-
priest of Anu and Antu of the Resh sanctuary, scribe of (the
series) Enūma-Anu-Enlil, from Uruk". The "series" mentioned
is the famous series of astronomical omens mentioned in the initial
sections of this chapter. The second family has Sin-leqē-unninī
as its ancestor, "scribe of Enūma-Anu-Enlil, kalū-priest of Anu
and Antu, from Uruk". Both these "ancestors" are known from
colophons of other tablets of the Seleucid period and the question
to what extent these scribal families were real families or merely
scribal schools seems undecided. Also the significance of owner-

ship and scribe escapes us. We do not know, e. g., whether the "scribe" of an ephemeris was its actual computer or not. All that one can safely say is that our tablets came from "priestly" circles, but this says little more than the trivial statement that they were written by professional scribes. And no information about the origin of these methods can be obtained from the colophons of the Uruk tablets.

The Babylon texts give us still less information about their scribes. From Pliny, Strabo, and Vettius Valens, however, are known names of three Babylonian astronomers who seemed also to appear in the colophons of our texts. One, Sudines, seemed to be contained in the second half of Anu-ahhē-šu-iddina, but the latter turned out to be a misreading of Anu-aha-ušabšī. The second name, Naburianos, seems to be attested once, in doubtful context in one of the very latest tablets, in the form Naburimannu but the reading itself is not really certain. And there is still less proof that Naburianos is mentioned as the inventor of the lunar System A to which the text belongs. Finally there is the name of Kidenas which corresponds to cuneiform Kidinnu. This name appears in a few colophons in the connection "tersitu of Kidinnu" which was guessed to mean "lunar tablet of Kidinnu" or "system of Kidinnu" and thus Kidinnu is usually considered to be the inventor of System B. This may be so, but real proof is missing. The term tersitu is a complete puzzle in this context because it is otherwise known to denote some tools or ingredients in connection with the manufacturing of glazed bricks.

Attempts were made to give accurate dates for the invention of the lunar theory. They were based on a comparison of modern computation with the results of the ancient theory. The slowly accumulating error of the ancient theory was supposed to be zero at the beginning, and this led to the alleged date of the alleged inventors Naburianu and Kidinnu. It suffices to say that this method presupposes the accuracy of the initial values, a hypothesis which is far from even being plausible. It is furthermore assumed that the parameters used in the actual computation of the ephemerides are exactly identical with the empirical values or, at least, with the values theoretically abstracted as correct from some observations. But we have seen that the parameters of the ephemerides were adjusted for the purpose of convenient computation. The errors caused by this procedure are very small; nevertheless they influence quite essentially the results of computations which are themselves based on the investigation of the small deviation from the factual motions. Hence there is no hope of obtaining, in this way, accurate information as to the date of invention of mathematical astronomy. For the time being, we must be satisfied with general historical considerations, however inconclusive they may appear. Otherwise one can only hope that a tablet may be found (and perhaps even published) which gives us direct information about the theoretical and empirical foundations of the whole theory.

BIBLIOGRAPHY TO CHAPTER V

All texts known to me by 1955 which concern mathematical astronomy are published in O. Neugebauer, Astronomical Cuneiform Texts, London, Lund Humphries, 1955 (3 vols.). This edition contains complete transcriptions, translations, and commentaries. Henceforth quoted as ACT.

Copies of all other available texts from the British Museum with much information about unpublished material are given in T. G. Pinches–J. N. Strassmaier–A. J. Sachs, Late Babylonian Astronomical and Related Texts, Providence, Brown University Press, 1955.

For a modern comprehensive discussion of the role of divination and astrology see J. C. Gadd, Ideas of Divine Rule in the Ancient East, The Schweich Lectures of the British Academy 1945 (London 1948).

The reader should be warned against the use of Jeremias, Handbuch der altorientalischen Geisteskultur. With the use of an enormous learned apparatus, the author develops the 'panbabylonistic" doctrine which flourished in Germany between 1900 and 1914, only to be given up completely after the first world war. The main thesis of this school was built on wild theories about the great age of Babylonian astronomy, combined with an alleged Babylonian "Weltanschauung" based on a parallelism between "macrocosm and microcosm". There was no phenomenon in classical cosmogony, religion, literature which was not traced back to this hypothetical cosmic philosophy of the Babylonians. A supreme disregard for textual evidence, wide use of secondary sources and antiquated translations, combined with a preconceived chronology of Babylonian civilization, created a fantastic picture which exercised (and still exercises) a great influence on the literature concerning Babylonia. Kugler was one of the few scholars in Germany who did not fall for these theories. In a little book called "Im Bannkreis Babels" he demonstrated drastically the absurdities which can be reached by the panbabylonistic methods. He collected 17 pages of striking parallels between the history of Louis IX of France and Gilgamesh, showing that Louis IX was actually a Babylonian solar hero.

The panbabylonistic school no longer has any followers. But it seems to me that Kugler's example should be studied by every historian because it demonstrates far beyond its original purpose how easy it is to fit a large body of evidence into whatever theory one has decided upon.

NOTES AND REFERENCES TO CHAPTER V

ad 42. The latest dated cuneiform tablet (75 A.D., thus the time of Vespasian) is an "Almanac" in the classification of Sachs (J. Cuneiform Studies 2, 1948, p. 280). It was found in Dropsie College in Philadelphia and in all probability came from Babylon. The exact date was established by Schaumberger.

ad 44. The story of the text from the "Frau Professor Hilprecht Collection of Babylonian Antiquities im Eigentum der Universität Jena" (this title is accurate) is somewhat peculiar, though not unique. Six lines from the reverse were published in 1908 in the Sunday supplement of the Münchner Neueste Nachrichten. Rather fantastic interpretations were made on the basis of this excerpt until Thureau-Dangin in 1931 suggested the explanation of the numbers as meaning distances in depth. During all these years the text could not be checked because it was "lost". In 1931, however, I got permission to have access to the closely guarded Jena collection, whose rich material, incidentally, furnished me with the key to the understanding of the relationship between multiplication tables and division (cf. above p. 31 f.). In going through this collection I found a tablet with the label "One of the 5 important Nippur texts from my desk drawer"; this turned out to be the lost text. Shortly thereafter, I was informed by the authorities in Jena that it was only by mistake that I had been admitted to the collection and that I was forbidden to publish any text from this collection. Nevertheless I reserved for myself the privilege of remembering my newly acquired knowledge, and since then my copy of the text has been used by other scholars. The essential passages are discussed in a review in Quellen und Studien zur Geschichte d. Math., Ser. B, vol. 3 p. 273 ff. (1936).

For the Old-Babylonian observations of Venus see Langdon-Fotheringham-Schoch, The Venus Tablets of Ammizaduga, Oxford, 1928. The chronological conclusions of this work have been disproved by subsequent archaeological evidence.

A comprehensive study of the series "Enūma Anu Enlil" was begun by E. F. Weidner in the Archiv für Orientforschung, vol. 14 (1942) p. 172–195, 308–318 and vol. 17 (1954) p. 71–89. The reader will find in this paper a description of the very complex structure of this "series" with its supplementary series of excerpts, commentaries, etc. It is important to realize that we have only very few original texts with astrological omina from a period before late Assyrian and Neo-Babylonian times. Consequently even the history of this early stage of astrological literature must largely be reconstructed from much later documents.

For the two tablets of the series "mul Apin" see Bezold-Kopff-Boll, Zenit- und Aequatorialgestirne am babylonischen Fixsternhimmel, Sitzungsberichte d. Heidelberger Akad. d. Wiss., phil.-hist. Kl., 1913, No. 11, and

E. F. Weidner, Ein babylonisches Kompendium der Himmelskunde, Am. J. of Semitic Languages and Literatures 40 (1924) p. 186–208. Cf. furthermore B. L. van der Waerden, Babylonian astronomy II. The thirty-six stars. J. of Near Eastern Studies 8 (1949) p. 6–26; also part III: The earliest astronomical computations (ibid. 10, 1951, p. 20–34).

The number of attested leap years in Babylonian texts is now sufficiently great to show that the 19-year cycle was introduced into consistent calendaric use very close to 380 B.C. This gives Meton's announcement of the cycle in Athens a priority of about 50 years and opens the possibility of an originally Greek discovery. On the other hand, A. Sachs has put forward (J. Cuneiform Studies 6, 1952, p. 113) arguments which connect the Babylonian intercalation rules with observations of the heliacal risings of Sirius prior to 380 B.C.

It may be remarked in this connection that "year" in astronomical context always means sidereal years in Babylonian texts (cf. Neugebauer ACT p. 70) but that there is no reason for assuming that one realized the difference between sidereal, tropical, and anomalistic year. Apparently it was Ptolemy who first defined "year" as meaning "tropical year" (cf. Almagest III, 1 p. 192 f. Heiberg).

In recent years, it has become commonplace (e. g., the first edition of this book or van der Waerden in Archiv für Orientforschung 16, 1953, p. 22) to consider 419 B.C. as the earliest attested date for the mention of the real zodiacal signs in Babylonia. A copy of the text in question, an astronomical diary for the year −418/417, has been published by E. F. Weidner in Archiv für Orientforschung 16 (Pl. XVIII) and shows, on the contrary, that zodiacal signs had not yet been introduced. Four passages occur (obv. 7, 11, rev. 8, 11 f.) where planets are said to be "behind" or "in front of" the alleged zodiacal signs. From this it is clear that ecliptical constellations, not zodiacal signs, are referred to [A. Sachs].

ad 45. Budge, The Rise and Progress of Assyriology, London 1925, writes about Strassmaier as follows (p. 228): "He was convinced that it was a waste of time to compile an Assyrian Dictionary, or to write a history of the Sumerian and Babylonian civilizations, whilst so many tens of thousands of tablets in the British Museum and elsewhere remained unpublished". Today one may repeat this statement, only replacing "tens of thousands" by "hundreds of thousands".

ad 46. The "observational" texts are discussed by A. Sachs, A Classification of Babylonian Astronomical Tablets of the Seleucid Period. J. Cuneiform Studies 2 (1950) p. 271–290.

ad 49. As an example of the solar motion according to System A, I have computed the data for the same year which we used on p. 110 for System B. These elements are readily obtainable from ephemerides slightly earlier or slightly later. As a matter of fact, all texts of System A form a uniform ephemeris with no disturbances at all from the earliest to the latest text known. This is not the case for System B and therefore every comparison between the two systems must reckon with the possibility that the texts of System B show some small individual deviations. Nevertheless it is clear from the difference in method that deviations of about 2° in solar longitude may occur in the column B. This does not imply, however, that the final columns show equally large deviations.

The ephemeris for System A leads to the following values

	XII₂	22,18,45	♈
2,59	I	20,26,15	♉
	II	18,33,45	♊
	III	16,41,15	♋
	IV	14,48,45	♌
	V	12,56,15	♍
	VI	12,56	♎
	VII	12,56	♏
	VIII	12,56	♐
	IX	12,56	♑
	X	12,56	♒
	XI	12,56	♓
	XII	11,56,15	♈

The dotted lines indicate the discontinuities at ♍ 13 and ♓ 27 where the monthly solar velocity changes from 28;7,30° to 30° and vice versa. The agreement with System B is quite close for the last month of the year. For 2,58 XII₂ System B gave ♈ 22;8,18,16 as compared with ♈ 22;18,45. For 2,59 XII we had ♈ 11;30,48,56 against ♈ 11;56,15 now. But for the middle of the year the linear zigzag function leads to ♍ 14;58,18,18 as compared with ♍ 12;56,15 just before the discontinuity.

ad. 50. O. Neugebauer, "Saros" and lunar velocity in Babylonian astronomy. Kgl. Danske Vid. Selsk., mat.–fys. medd. 31,4 (1957).

ad 51. The "Saros". It has become customary to call the relation

223 synodic months = 242 draconitic months

the "Babylonian Saros" and to assume that it was the basis for the prediction of eclipses by the Babylonians and their successors.

Unsuccessful protests against the use of this terminology have been made by Ideler (1825[1])), Tannery (1893[2])), Schiaparelli (1908[3])), Bigourdan (1911[4])), and Pannekoek (1917[5])). Only Ideler, however, gave an account of the origin of this term, and it therefore seems to me worthwhile to outline the main steps as a beautiful example of the creation of generally accepted historical myths.

The Sumerian sign šár has, among others, the meaning "universe" or the like. As a number word it represents 3600, thus being an example of the transformation from a general concept of plurality to a concrete high numeral. In the special meaning of 3600 years, "Saros" is used by Berossos[6]) (about 290 B.C.) and, following him, by Abydenus[7]) (second cent. A.D.) and by Synkellos[8]) (about 800 A.D.).

[1]) Handbuch der ... Chronologie I p. 213.
[2]) Recherches sur l'histoire de l'astron. anc. p. 317
[3]) Scritti I p. 75.
[4]) L'astronomie p. 33.
[5]) The origin of the Saros. Koninklijke Akad. van Wetenschappen te Amsterdam, Proceedings 20 (1917) p. 943–955. Cf. also Quellen und Studien z. Gesch. d. Math., Ser. B. vol. 4 (1937–1938) p. 241 ff. and p. 407 ff.
[6]) Fragm. 29 ff. (Schnabel, Berossos, p. 261 ff.).
[7]) Schnabel, Berossos, p. 263, 30 a (in line 29, correct καὶ, γ to read καὶ ,γ).
[8]) Schnabel, Berossos, p. 261 ff.

An astronomical meaning is associated with "Saros" for the first time in
the encyclopedia of Suidas (about 1000 A.D.). There "Saros" is explained as
"a measure or number with the Chaldeans" and then the remark is added
that one Saros contains 222 months, i. e., 18 years and 6 months, while 120
Saroi correspond to 2222 years[1]). The first relation implies that one year contains
exactly 12 months. This excludes the Babylonian calendar. The second relation
is a consequence of the first if one considers 2222 to be a scribal error for 2220;
otherwise it is senseless. In no case is there any relation to eclipses.

Pliny, NH II, 56[2]), discusses the recurrence of eclipses after 223 months.
The manuscripts contain different readings of this number: 213 or 293 or 222
or 235[3]). Edmund Halley had at his disposal a Pliny text which gave 222. He
realized that only 223 made sense and assumed that a similar correction should
be made in Suidas, whose source he thought to be Pliny. He overlooked,
however, that all the other figures in Suidas contradict the change of 222 into
223 and that they are only the expression of the trivial relation that one year
contains 12 months. Thus Halley assumed that Suidas intended to say that
223 months were called one "Saros", and he published this conjecture in the
Philosophical Transactions 1691 (p. 535–540; reprinted in the Acta Eruditorum
1692, p. 529–534).

Halley's hypothesis was severely criticized by Le Gentil in 1766[4]) after it
had been presented as a fact by Montucla in the first edition of his Histoire
des mathématiques (1758). A more cautious formulation in the second edition
(1802) was too late to have any effect. Since Montucla it has been an accepted
doctrine of textbooks that the Babylonians used the "saros" for the prediction
of eclipses. From Kugler (1900) we know how eclipses were actually computed
during Seleucid period, namely, by a careful investigation of the latitude of the
moon in relation to the syzygies. Nevertheless, there are certain indications that
the periodic recurrence of lunar eclipses was utilized in the preceding period
by means of a crude 18-year cycle which was also used for other lunar phenomena.

The myth of the Saros is often used as an "explanation" of the alleged
prediction by Thales of the solar eclipse of –584 May 28. There exists no cycle
for solar eclipses visible at a given place; all modern cycles concern the earth
as a whole. No Babylonian theory for predicting a solar eclipse existed at 600
B.C., as one can see from the very unsatisfactory situation 400 years later; nor
did the Babylonians ever develop any theory which took the influence of geographical latitude into account. One can safely say that the story about Thales's
predicting a solar eclipse is no more reliable than the other story about An-
axagoras predicting the fall of meteors.

Even from a purely historical viewpoint the whole story appears very doubtful.
Our earliest source, Herodotus (I, 74), reports that Thales had predicted "this
loss of daylight" to the Ionians correctly "for the year" in which it actually
happened. This whole formulation is so exceedingly vague that in itself it excludes
the use of any exact method. The farther we move away from the time of Thales,

[1]) Ed. A. Adler IV p. 329.
[2]) Ed. Ian-Mayhoff I p. 144; Loeb Class. Libr. I p. 204/205.
[3]) Cf. the critical apparatus in Ian-Mayhoff. The number 235 was obviously
suggested by the 19-year "Metonic" cycle which contains 235 months.
[4]) Mémoires for 1766 of the Acad. Royale des Sci., Paris, p. 55 ff.

the more generous do the ancient authors become in assigning to him mathematical and astronomical discoveries. I see not a single reliable element in any of these stories which have become so dear to the histories of science (cf. also p. 148). In this connection may be quoted the summary of an article by R. M. Cook, Ionia and Greece, 800–600 B.C., J. Hellenic Studies 66 (1946) p. 67–98: "My tentative conclusion is that we do not know enough to say definitely whether in the 8th and 7th centuries the Ionians were generally the pioneers of Greek progress, but that on the present evidence it is at least as probable that they were not."

Such a cautious outlook is, however, far from common. Only one of the amusing fairy tales, spun out of the story of the Thales eclipse, may be mentioned. In a learned work on "Die griechisch-römische Buchbeschreibung verglichen mit der des vorderen Orients" (Halle 1949) C. Wendel declares (p. 20 ff.) that the "Bahnbrecher" of Ionian science must have had a library of "amazing richness" at their disposal and Thales must have been its "spiritual founder". Needless to say Thales's studies in Egypt are also taken very seriously. Unfortunately, we know from Diodorus I, 38 that Thales knew so little about Egypt that he could propose the theory that the inundation of the Nile began at "the mouths of the river when the etesian winds . . . hinder the flow of the water into the sea".

ad 52. The problem of column Q can be easily formulated as follows. Column O establishes the elongation $\Delta\lambda$ of the moon from the sun for the critical evening by multiplying the time since conjunction by the relative velocity between sunset and moonset. This is equivalent to asking for the time it takes the arc $\Delta\lambda$ of the ecliptic to set and this in turn is equal to the time of rising of the diametrically opposite section of the ecliptic. But for the rising times of ecliptic arcs, arithmetical schemes are known (cf. below p. 159 and Fig. 26) which solve the problem of transforming longitudes λ into right ascension α. Exactly the same coefficients appear in the transformation from $O = \Delta\lambda$ into $Q = \Delta\alpha$ as I have shown in J. Cuneiform Studies 7 (1953) p. 100 to 102.

ad 56. For the schematic computation of the dates of the appearances (Γ), oppositions (Θ), and disappearances (Ω) of Sirius cf. A. Sachs, Sirius Dates in Babylonian Astronomical Texts of the Seleucid Period, J. Cuneiform Studies 6 (1952) p. 105–114.

ad 57. O. Neugebauer, The Babylonian method for the computation of the last visibility of Mercury. Proc. Am. Philos. Soc. 95 (1951) No. 2.

ad 58. For Babylonian planetary theory cf. vol. 2 of my Astronomical Cuneiform Texts. How these methods originated is still largely unknown. A. Sachs has given a classification of the not strictly mathematical texts (J. Cuneiform Studies 2, 1950, p. 271–290) which also clearly distinguish between observational elements and computed data, e. g. on the basis of periodic recurrence.

Apart from these well defined classes, there exist procedure texts which evidently belong to an intermediate state preliminary to the final mathematical theory. Texts of this type still present serious difficulties because of unknown terminology, composite character of historical background, etc. Several copies of such texts have been published by Thureau-Dangin in Tablettes d'Uruk (Paris 1922) and in Pinches–Strassmaier–Sachs (cf. p. 138). So far

Sachs and I have succeeded only in explaining the meaning of one of these texts to a reasonable degree of certainty (J. Cuneiform Studies 10, 1956, p. 131–136).

ad 59. The secrecy of ancient oriental sciences has often been assumed without any attempt to investigate the foundation for such a hypothesis. There exist indeed examples of "cryptographic" writing both in Egypt and Mesopotamia. The Cenotaph of Seti I, for instance, contains cryptographic passages in the mythological inscriptions which are written around the sky goddess. Some of the passages use rare readings of hieroglyphs, some are simply incorrectly arranged lines of the original from which the artist copied. In a related text concerning a sun dial the words are written backwards, as if reflected by a mirror, but this causes no real difficulty in reading the text. On the whole, however, all texts with mathematical or astronomical context show not the slightest intention of concealing their meaning from the reader. I think one can only agree with T. E. Peet, the editor of the mathematical Papyrus Rhind, that we have no reason for assuming the existence of any secret science in Egypt.

The same holds for Babylonia. The Old Babylonian mathematical texts are as plainly written as possible. From the latest period there exist a few texts which give lists of numbers and signs obviously for coding and decoding purposes. A few words and proper names are written in such a code in the colophons of two ephemerides. The ephemerides themselves as well as the procedure texts show no trace of an attempt to hide their contents. If many details remain unintelligible to us, it is our ignorance and missing texts which cause the difficulties, not an intentionally cryptic writing. I think the remark found occasionally in colophons of Uruk texts that the text should only be shown to "the informed" is not to be taken too seriously. It hardly indicates much more than professional pride and feeling of importance of members of the scribal guild.

Cryptographic devices occur also in Greek manuscripts of the Byzantine period, based, e. g., on a simple substitution of letters and their numerical values in inverse order; cf. e. g., V. Gardthausen, Griechische Palaeographie II (2nd ed., Leipzig 1913) p. 300 ff. Magical texts are, of course, full of secret combinations of letters; astrological texts, however, are practically free of such secrecy.

The sixth chapter of the Sūrya-Siddhānta deals with a graphical representation of the different phases of an eclipse. It ends with the remark, "This mystery of the gods is not to be imparted indiscriminately: it is to be made known to the welltried pupil, who remains a year under instruction". Burgess says rightly, "It seems a little curious to find a matter of so subordinate consequence ... guarded so cautiously ...". The same holds for the construction of a celestial globe (S.-S. XIII, 17 and similarly Pañca-S. XIV, 28). Similarly one of the most trivial chapters in the Pañca-Siddhāntika (XV) is called the "secrets of astronomy". Here one gets the impression that we are dealing with a very old section which could have been omitted without any harm to the understanding of the rest.

The name of Kidenas = Kidinnu is customarily associated with the city of Sippar and its school of astronomers, mentioned by the Greek writers cited above p. 137. A. Sachs has realized, however, that the passage in question was misread by Strassmaier (cf. Neugebauer, ACT I p. 22 colophon Zo). We have not a single astronomical text which came from Sippar.

CHAPTER VI

Origin and Transmission of Hellenistic Science.

60. Any attempt to reconstruct the origin of Hellenistic mathematics and astronomy must face the fact that Euclid's "Elements" and Ptolemy's "Almagest" reduced all their predecessors to objects of mere "historical interest" with little chance of survival. As Hilbert once expressed it, the importance of a scientific work can be measured by the number of previous publications it makes superfluous to read.

Because Euclid's work falls not much more than a century after the beginning of scientific mathematics, it has been easier to restore its prehistory than is the case with astronomy. The early date of Euclid (about 300 B.C.) leaves room for two or more centuries of active development carried out by men like Archimedes and Apollonius. Ptolemy, in 150 A.D., lives close to the end of the Hellenistic age, and his work comprises practically all astronomical achievements which could be reached with the mathematical methods of antiquity. The careful analysis, on purely mathematical grounds, of Euclid's work has given valuable information about the preceding major steps on which it was built. Ptolemy's work is exclusively concerned with the description of one unified method for the representation of the celestial phenomena. On the basis of the Almagest we would have no idea about the existence of totally different methods, Greek and Oriental, which preceded and occasionally even survived the Almagest.

Finally one must realize that the "Elements" of Euclid concern, with very few exceptions, a purely Greek development in a sharply defined direction. Ptolemy's astronomy is probably built to a large extent on results obtained 300 years earlier by Hipparchus, who in turn was influenced both by Greek and by Babylonian ideas. Hence the problems connected with the history of astronomy are far more involved than is the case with mathematics. Furthermore Greek mathematical procedures are directly intelligible to a modern mathematician whereas ancient astronomical treatises operate with a terminology and with problems and empirical and numerical methods which are no longer familiar in our time. This situation is also reflected in the modern discussion of these problems. For the history of Greek mathematics, there are quite competent and complete presentations but we are far from this goal in the history of ancient astronomy.

61. To say that Greek mathematics of the Euclidean style is a strictly Greek development does not mean to deny a general Oriental background for Greek mathematics as a whole. Indeed, mathematics of the Hellenistic period, and still more of the later periods, is in part only a link in an unbroken tradition which reaches from the earliest periods of ancient history down to the beginning of modern times. As a particularly drastic example might be mentioned the elementary geometry represented in the Hellenistic period in writings which go under the name of Heron of Alexandria (second half of first century A.D.). These treatises on geometry were sometimes considered to be signs of the decline of Greek mathematics, and this would indeed be the case if one had to consider them as the descendants of the works of Archimedes or Apollonius. But such a comparison is unjust. In view of our recently gained knowledge of Babylonian texts, Heron's geometry must be considered merely a Hellenistic form of a general oriental tradition. The fact, e. g., that Heron adds areas and line segments can no longer be viewed as a novel sign of the rapid degeneration of the so-called Greek spirit, but simply reflects the algebraic or arithmetic tradition of Mesopotamia. On this more elementary level, the axiomatic school of mathematics had as little influence as it has today on surveying. Consequently, parts of Heron's writings, practically unchanged, survived the destruction of scientific mathematics in late antiquity. Whole sections from these works are found again, centuries later, in one of the first Arabic mathematical works, the famous "Algebra" of al-Khwārizmī (about 800 to 850). This relationship can be especially easily demonstrated by means of the figures. In order to make the examples come out in nice numbers, the figures were composed from a few standard right triangles. One of these

standard examples is shown in Fig. 21, which appears in Heron as well as in al-Khwārizmī. Two right triangles with sides 8, 6, and 10 are combined into an isosceles triangle of altitude 8 and base 12. Then a square should be inscribed. The resulting linear equation yields for the side of the square 4 $\overline{2}$ $\overline{5}$ $\overline{10}$, i. e. 4⅘. The style of the formulation of these problems, the way of solving them in special numerical examples—all this closely resembles the Babylonian mathematical texts. A similar comparison could

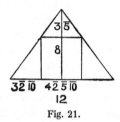

Fig. 21.

be carried out for various parts of Hellenistic and Arabic mathematics, such as the inheritance problems, the algebra of the Diophantine type, etc. This does not mean that Hellenistic or even Arabic authors were able to utilize Babylonian material directly. All that we can safely say is that a continuous tradition must have existed, connecting Mesopotamian mathematics of the Hellenistic period with contemporary Semitic (Aramaic) and Greek writers and finally with the Hindu and Islamic mathematicians.

62. The question arises whether any Oriental influence is apparent in the scientific branch of Greek mathematics. My answer to this question cannot be proved by documentary evidence, but the following working hypothesis seems to me to account for the known facts: the theory of irrational quantities and the related theory of integration are of purely Greek origin, but the contents of the "geometrical algebra" utilize results known in Mesopotamia.

To substantiate these statements a few remarks must be made about the historical development of Greek mathematics. First of all, it seems necessary to distinguish sharply between the axiomatic style of mathematics, which is the work of Eudoxus and his contemporaries in the fourth century B.C., and the

mathematics usually connected with the Ionian and South-
Italian schools. I see no reason to deny to the earlier period a
comparatively large amount of mathematical knowledge which
might comprise, or even exceed, in certain points the knowledge
attested in Mesopotamian sources. It seems to me evident, however, that the traditional stories of discoveries made by Thales
or Pythagoras must be discarded as totally unhistorical. Thales,
e. g., is credited with having discovered that the area of a circle
is divided into two equal parts by a diameter. This story clearly
reflects the attitude of a much more advanced period when it
had become clear that facts of this type require a proof before
they can be utilized for subsequent theorems. To the later mathematicians it seemed natural to assume that theorems which had to
be established first on logical grounds should also come first chro-
nologically. Actually the Greek historians acted in exactly the same
way as modern historians do when no source material is available to them: they restored the sequence of events according to
the requirements of the theory of their own times. We know today
that all the factual mathematical knowledge which is ascribed
to the early Greek philosophers was known many centuries before,
though without the accompanying evidence of any formal method
which the mathematicians of the fourth century would have
called a proof. For us, there is nothing to do but to admit that
we have no idea of the role which the traditional heroes of Greek
science played. It seems to me characteristic, however, that
Archytas of Tarentum could make the statement that not geometry
but arithmetic alone could provide satisfactory proofs. If this was
the opinion of a leading mathematician of the generation just
preceding the birth of the axiomatic method, then it is rather
obvious that early Greek mathematics cannot have been very
different from the Heronic Diophantine type.

It is also generally accepted that the essential turn in the
development came about through the discussion of the consequences of the arithmetical fact that no ratio of two integers could
be found such that its square had the value 2. The geometrical
corollary that the diagonal of a square could not be "measured" by
its side obviously caused serious discussion about the relation
between geometrical and arithmetical proof. The "paradoxa"
concerning continuity, both of space and time, made the relation

to the whole problem of determination of area and volume evident. One way out might have been the assumption of a somewhat atomistic structure of geometrical objects by means of which the problem of area or volume would have been reduced, though in not too clear a fashion, to a counting of discrete elements, "atoms".

The reaction of the mathematicians against this type of speculation seems to have led to two major steps. First of all, one had to agree exactly on a system of basic assumptions from which alone the rest had to be deduced; this gave rise to the strictly axiomatic procedure. Secondly, it had become clear that one should consider the geometrical objects as the given entities such that the case of integer ratios appeared as a special case of only secondary interest; this led to the problem of how to formulate classical arithmetical and algebraic knowledge in geometrical language. The result is the familiar "geometrical algebra" of Greek mathematics. It is these two essential steps which are fully to the credit of the Greek mathematicians.

The situation changes when we ask about the origin of mathematical relations which were incorporated in the systematic building of geometrically demonstrated laws. Everything which is directly related to the theory and classification of irrational quantities is, of course, Greek; and the same holds for the rigid theory of the processes of integration. The elementary theory of numbers, however, may or may not eventually be based on much older oriental material. I do not doubt that any connection with the name of Pythagoras is purely legendary and of no historical value.

The most interesting question, however, seems to me the problem of the origin of the "geometrical algebra". We have seen that the Babylonian treatment of problems of second degree consists in reducing them to the "normal form" where two quantities, x and y, should be found from their given product and their sum or difference. It seems significant that the geometric formulation of this problem leads precisely to the central problem of the geometrical algebra, a problem which is otherwise rather difficult to motivate. This problem is known as the "application of area", which consists, in its simplest form, in the following. Given an area A and a line segment b; construct a rectangle

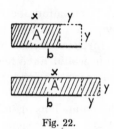

Fig. 22.

of area A such that one of its sides falls on b but in such a way (cf. Fig. 22) that the rectangle of equal height and of length b is either larger or smaller by a square than the rectangle of area A. The identity of this strange geometrical problem with the Babylonian "normal form" is at once evident when we formulate it algebraically. Let us call, in both cases, x and y the sides of the rectangle. Then we are given

$$xy = A.$$

In the first case a square should remain free; its sides are y and we must require

$$x + y = b.$$

In the second case, a square should exceed the rectangle of side b; thus we should have

$$x - y = b.$$

These are indeed the two normal forms (cf. p. 41).

Attempts have been made to motivate the problem of "application of areas" independently of this algebraic background. There is no doubt, however, that the above assumption of a direct geometrical interpretation of the normal form of quadratic equations is by far the most simple and direct explanation. I realize that simplicity is by no means equivalent with historical proof. Nevertheless the least one must admit is the possibility of the above explanation. That the numerical solution of quadratic equations was taught in contemporary Babylonian scribal schools cannot be doubted in view of the fact that this is still attested even in later periods of cuneiform writing. The only serious question consists in the specific way in which such knowledge found its way to Greece. Here we are left to mere speculation.

But it seems to me not to require too much imagination to think of a diffusion of mathematical knowledge from the Near East to Greece in a period close to the eve of the Macedonian offensive against the Persian empire.

In the history of speculative thought much has been said about the direct contact of Plato and Aristotle with Orientals. An Iranian is reported to have informed Plato about the religion of Zarathustra. Callisthenes, the nephew of Aristotle, is even supposed to have brought Babylonian astronomical records to Athens. This latter report is not too trustworthy, as T. H. Martin has rightly emphasized in 1864. No mention of it comes directly from Aristotle but it is based only on Porphyrius (third cent. A.D.) through Simplicius (sixth cent. A.D.); and worst of all, these Babylonian observations are supposed to have reached 31,000 years back. Whatever the case may be, a few years later Babylonia was under Greek domination and literary evidence is then no longer required to prove that the Greeks had access to Babylonian science from this time onwards. Nevertheless, the need for caution remains when the contact under discussion especially concerns the work of Eudoxus. I see no good reason to deny the possibility of his travels to Egypt. It seems to me certain, however, that there was nothing to learn from the Egyptians themselves, and the hypothesis that Babylonian science had reached Egypt before Greece seems to me only to substitute one name of an unknown quantity for another. At the present state of our knowledge, none of these stories contributes significantly to our insight into the historical events.

63. The Greeks themselves had many theories about the origin of mathematics. A favored one, which is still kept alive in modern textbooks, makes the necessity of repeated land measurement responsible for geometry. Modern authors have often referred to the marvels of Egyptian architecture, though without ever mentioning a concrete problem of statics solvable by the known Egyptian arithmetical procedures. A much more sophisticated attitude is represented by Aristotle, who considers the existence of a "leisure class", to use a modern term, a necessary condition for scientific work. Our factual knowledge about the development of scientific thought and of the social position of the men who were responsible for it is so utterly fragmentary, however, that it seems to me completely impossible to test any such hypothesis, however plausible it may appear to a modern man.

It seems to me equally impossible to give any one conclusive "explanation" for the origin of higher mathematics in the fifth and fourth century in Athens and the Italian colonies. On the negative side, however, I think that it is evident that Plato's role has been widely exaggerated. His own direct contributions to mathematical knowledge were obviously nil. That, for a short while, mathematicians of the rank of Eudoxus belonged to his circle is no proof of Plato's influence on mathematical research. The exceedingly elementary character of the examples of mathematical procedures quoted by Plato and Aristotle give no support to the hypothesis that Theaetetus or Eudoxus had anything to learn from Plato. The often adopted notion that Plato "directed" research fortunately is not borne out by the facts. His advice to the astronomers to replace observations by speculation would have destroyed one of the most important contributions of the Greeks to the exact sciences. Plato's doctrines undoubtedly have had great influence upon the modern interpretation of Greek sciences. But if modern scholars had devoted as much attention to Galen or Ptolemy as they did to Plato and his followers, they would have come to quite different results and they would not have invented the myth about the remarkable quality of the so-called Greek mind to develop scientific theories without resorting to experiments or empirical tests.

64. The structure of our planetary system is indeed such that Rheticus could say "the planets show again and again all the phenomena which God desired to be seen from the earth". The investigations of Hill and Poincaré have demonstrated that only slightly different initial conditions would have caused the moon to travel around the earth in a curve of the general shape given in Fig. 23, and with a speed exceedingly low in the outermost quadratures Q_1 and Q_6 as compared with the motion at new and full moon. Nobody would have had the idea that the moon could rotate on a circle around the earth and all philosophers would have declared it as a logical necessity that a moon shows six half moons between two full moons. And what could have happened with our concepts of time if we were members of a double-star system (perhaps with some uneven distribution of

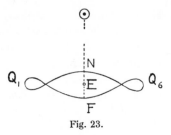

Fig. 23.

mass in our little satellite) is something that may be left to the imagination.

Actually, however, the initial conditions of our planetary system were chosen in such a way that all the satellites of the sun—and our own satellite as well—behave with great modesty. Their orbits can be closely approximated by circles such that the simplest possible model of a circular motion with constant speed leads immediately to very reasonable results for the description of the solar and lunar phenomena. On the other hand, the deviations from the trivial circular orbits are just great enough to be observed and to challenge an explanation, but small enough such that again comparatively simple modifications of the trivial solution give satisfactory results. The successive approximations of the Babylonian lunar and planetary theory reflect this situation perfectly. At the basis lies the counting of the periodically recurrent phenomena; the properly chosen periodic functions—zigzag functions or step functions—suffice to describe the deviation from a trivial mean motion.

Perhaps a little before these methods were developed in Mesopotamia, perhaps almost simultaneously, a most decisive step in another direction was made by Eudoxus. The then recent discovery of the sphericity of the earth must have suggested a corresponding sphericity of the sky and a circular motion of the celestial bodies. Eudoxus's theory may have well started from the following consideration. The motion of sun and moon can be described as the combination of the uniform motions of two concentric spheres: one is the fast daily rotation about the poles of the equator; the other is slow and proceeds in opposite direction about an inclined axis which is perpendicular to the ecliptic. Eudoxus saw that a similar combination is capable of explaining,

at least qualitatively, also the most striking phenomenon of
planetary motion, the retrogradations. The motion of two homocentric spheres allows of two trivial limiting types. If the two axes
are made to coincide, the body which is fixed to the equator of
one of the two spheres moves simply in a circle with the difference
velocity. If, secondly, the two opposite velocities are made of
equal amount the body remains stationary as seen from the center.

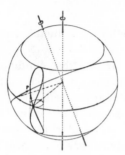

Fig. 24.

But one case remains to be investigated: what motion results from
equal opposite velocities but inclined axes? Eudoxus found that
the orbit is an 8-shaped curve (Fig. 24). Now one can superimpose
a third rotation about an axis which is perpendicular to the plane
of symmetry which represents the plane of the ecliptic. Consequently, the point P no longer follows a closed curve but proceeds
with a certain mean velocity in the ecliptic. Simultaneously,
however, there appears a periodic deviation from the ecliptic,
or a motion in latitude. Finally, one will obtain retrogradations
if the longitudinal component is less than the backward motion
in the original figure eight. Thus it is demonstrated that, at least
qualitatively, even the apparent irregularities of planetary motion
can be described by a combination of circular motions of uniform
angular velocity.

    In spite of the great importance, in principle, of the discovery
of Eudoxus, it is quite obvious that a model of this type has grave
shortcomings.

    For example, the observed retrogradations of the planets do
not recur in curves of identical shape as would be the case in
the Eudoxian model. Another difficulty lies in the large variation

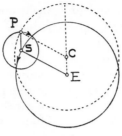

Fig. 25.

of the brightness of planets, that seemed to indicate corresponding variations in their distance from the earth. We do not know who first succeeded in explaining these and similar anomalies by means of a much more flexible modification of the theory of uniform circular motion. We know, however, that Apollonius (about 200 B.C.) used the simple device of viewing uniform circular motion, not from the center of the orbit, but from a slightly eccentric point. This obviously has the effect that the motion appears fastest where the circle is nearest to the observer and slowest at the opposite point. But Apollonius proved more. He demonstrated that an eccentric movement of this type can always be replaced by an epicyclic motion where the center of the epicycle moves on a circle with the observer at its center and with a radius of the epicycle equal to the eccentricity (cf. Fig. 25). All that is needed is to regulate the angular velocities in such a way that the point P and the observer E remain the vertices of a parallelogram $SPCE$. But as soon as epicycles are introduced, it is also clear that the motion of P around S can be chosen in such a way that P appears to have a retrograde motion if observed from E, as we have shown in the discussion of Fig. 13 b and 14 c (p. 123 f.). Hence the model of homocentric spheres could be abolished, and uniform description of all celestial motion was obtained by means of eccenters and epicycles. But the main principle, the fundamental role of circular motion, seemed to have been splendidly vindicated. This conviction remained the cornerstone of celestial "dynamics" of ancient astronomy comparable to a law of inertia.

In principle, however, ancient astronomers pretended only to "describe" the appearances, not to "explain" them. All that was actually observable was angular motions, the only exceptions being the distances for sun and moon obtainable by means of parallax. For the planets, however, neither theory nor observations were accurate enough to obtain reliable information as to their distances. In our discussion of the geocentric description of a heliocentric motion in Figs. 13 and 14, S represented the sun. Now we try only to describe the direction EP under which the planet P appears from E. Hence we can no longer say that S is the sun but only that ES is the direction to the sun. But otherwise all our conclusions remain valid. Thus we can say that the angular motion of an inner planet is described by an epicyclic motion such that the direction from E to the center S of the epicycle coincides with the direction from E to the sun. And an outer planet P moves on its epicycle around C in such a way that CP is always parallel to the direction from E to the sun. This is indeed the basic formulation of the Greek planetary theory by means of epicycles, with the obvious refinement that we should say "mean sun" instead of simply "sun". This theory is a correct description of the appearances so far as the angular motion is concerned and it would be a correct heliocentric theory if the correct scale were chosen. Second-order deviations from this first-order approximation could be explained by added eccentricities and similar devices which were brought to perfection by Ptolemy. Only greatly refined observations could eventually disclose the defects of the supposition of strictly circular motions.

65. If one looks back on this sketch of the development of the Ptolemaic planetary system, one sees no reason for the assumption of oriental influences. All that we know about Egyptian astronomy rules out any possible influence from this source. The Babylonian theory, on the other hand, is known to us to have reached about equally excellent results—by means of methods which nowhere point to an interpretation through a combination of circular motions or any other mechanical model. Indeed, zigzag and step functions practically exclude any such attempt. Nevertheless, Babylonian influence is visible in two different ways in Greek astronomy: first, in contributing basic empirical material for the geometrical theories which we have outlined in the preceding section; second, in a direct continuation of arithmetical methods which were used simultaneously with and independently of the geometrical methods.

The first influence was revealed when the Babylonian lunar theory was deciphered by Epping and Kugler. Exactly the same constants which determined the periods of several of the most important zigzag functions in the Babylonian theory are attested as the relations from which the mean motions were derived in the Greek theories, especially by Hipparchus (as nearly as we can tell from Ptolemy's references in the Almagest). Because the earliest Babylonian ephemerides antedate—though only by a narrow margin—the time of Hipparchus, it cannot be denied that at least the empirical foundations of the Babylonian theory must have been known to Hipparchus. How this knowledge was transmitted to him and how much he knew about the actual technique of computing ephemerides cannot be answered from our sources. Ordinarily, the teaching activity of the Babylonian Berossos (who moved to the Greek island of Cos about 270 B.C.) is considered responsible for the transmission of much of the astronomical knowledge to the Greeks. This may actually be the case, though the extant fragmentary excerpts from his writings contain no specific reference to mathematical astronomy. What we would really need in order to understand the details of transmission is a Greek commentary to Babylonian ephemerides and procedure texts. Somewhere the great step from year-by-year ephemerides to tables based on mean motions, as we know them from the Almagest, must have been made. That we cannot answer such a question even approximately demonstrates how little we know about the earlier period of Hellenistic astronomy outside Mesopotamia.

66. Babylonian influence is not restricted to providing essential constants for the determination of the parameters of the geometrical models of our planetary system. A much more direct development of the Babylonian arithmetical methods has become visible from Greek papyri and from occasional references to technical details in the astrological literature. Exactly as the "Greek" mathematical literature must be divided into two classes, a purely scientific development and a more elementary tradition which is closely related to the oriental tradition, so also do astronomical procedures fall into two groups: one leading to the Almagest, the other probably best known among the astrological authors for their computations of the positions of the celestial bodies for

horoscopic purposes. I call this second class of procedure the "arithmetical methods" or the "linear methods" because they are essentially based on difference sequences of first order.

It must be realized that no such classification is anything more than a convenient matter of speech and that there exist many contacts and influences between both extremes. Most of all, the reader should be warned not to take the expression "arithmetical methods" as implying that the methods of the Almagest somehow exclude numerical procedure. The opposite is true. Not only does the Almagest contain a great number of numerical tables, which in turn are based on an enormous amount of numerical computation, but the final goal of the Almagest is exactly the same as that of the "arithmetical methods", namely, to provide numerical data for astronomical phenomena. But the Almagest is unique in its desire to explain the empirical foundations and the theoretical reasons for its procedures. And the way always leads first to a definite geometrical model, from which the resulting arithmetical consequences are then derived. The linear methods, however, proceed on exclusively numerical grounds precisely in the fashion which is now familiar to us from the Babylonian texts. Though no theoretical treatise concerning these linear methods is preserved, it is clear that they rest on cleverly designed procedures and on empirical material which is probably rather similar for both methods. Nevertheless, the concept of a geometrical model seems to be completely absent for this second type of astronomical literature. This may be compared with the lack of a strictly axiomatic structure for the Heron-Diophantus type of Hellenistic mathematics.

The direct survival of Babylonian methods can be recognized most easily in an important problem of mathematical geography. In Hellenistic and medieval geography, one frequently finds the latitude of a locality expressed by means of the ratio of the longest to the shortest daylight for the region in question. Alexandria, e. g., falls in the zone for which this ratio is 7:5; that is to say, the longest daylight is 14 hours and the shortest night, assumed to be equal in length to the shortest day[1]), is 10 hours. Similarly,

---

[1]) This convenient assumption of exact symmetry was always made in ancient astronomy. Actually, however, atmospheric influences make the sun visible longer than would be the case with a mathematical horizon. Consequently the shortest daylight is longer than the shortest night.

Babylon is characterized by the ratio 3:2. Consequently it becomes an important problem to determine the length of daylight astronomically. Daylight will be longest when the sun is at the summer solstice, ♋ 0°. At sunrise on this day, the point ♋ 0° rises because the sun is located at this point. At sunset on this day, the sun and the point ♋ 0° set in the west, and the point ♑ 0° rises because, at any moment, 180° of the ecliptic must be above the horizon. Consequently, we can say that, during the longest day, the six signs ♋, ♌, ♍, ♎, ♏, ♐ have risen. Similarly, we can say that, during the shortest day, the six signs from ♑ to ♊ have risen.

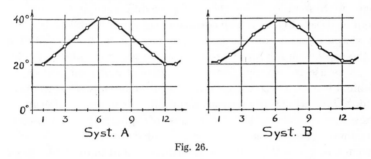

Fig. 26.

Thus we are able to determine the variation in the length of daylight if we know the time of rising of each individual zodiacal sign. If we call α_1 the rising time or "ascension" of ♈, α_2 of ♉, etc., then we know that the longest daylight is given by the sum $\alpha_4 + \ldots + \alpha_9$, and the shortest daylight by $\alpha_{10} + \ldots + \alpha_3$. In general, for any time of the year, we know the length of daylight if we know the rising times for the semicircle of the ecliptic which begins at the point in which the sun is located.

The values of the ascensions depend on the variable inclination between ecliptic and horizon and vary in a rather complicated way.[1] Nevertheless, arithmetical schemes to account for their variation were devised for the computation of column C (length of daylight) in the Babylonian lunar ephemerides (cf. p. 116 f.), such that the extremal values show the ratio 3:2 mentioned above as the characteristic ratio for the latitude of Babylon. These arithmetical schemes are slightly different for System A or B. In System A, the α's increase and decrease with constant differ-

---

[1] For an observer on the equator the rising times are called "right ascensions", a term which is still in use in modern astronomy.

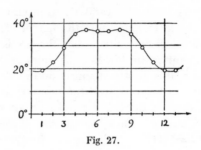

Fig. 27.

ence; in System B, twice the ordinary difference is assumed in the middle (cf. Fig. 26). The correct curve would look like Fig. 27. Its shape depends on the geographical latitude, the indentation at the top becoming more pronounced as we move north. Tables for these ascensions are given in all ancient and medieval astronomical works. These tables demonstrate concretely what we have said before about the classification of astronomical literature. They may be computed by means of spherical trigonometry; this then leads to values as represented in Fig. 27. This is the case, e. g., for the tables in the Almagest and in the later Greek and Arabic works of similar character. In many smaller treatises, however, one finds the ascensions for various geographical locations computed with schemes which are exactly of the type of System A or B, represented in Fig. 26. This is the case not only in the West but also in Indian astronomical texts (beginning with the sixth century A.D.) and the same holds for the majority of the astrological writings until the end of the Middle Ages.

It may seem very strange that primitive arithmetical schemes were used long after the correct trigonometrical solutions had been found and utilized for the computation of tables. And not only were trigonometric and arithmetic schemes coexistent, but the arithmetical devices maintained the duplicity of Systems A and B inherited from the Babylonian ephemerides. This is a nice example of the "conservatism" of the human race as a whole because this parallelism of equivalent methods is attested for Babylonians, Greeks, Romans, Jews, Christians, and Moslems alike.

But even for the limited field of the history of mathematics, the problem of the "ascensions" is of great interest. The careful investigation of early Greek spherical geometry, especially of Theodosius and Menelaos, has shown that this problem is one of the major goals of the whole theory. It is probably Menelaos (about 100 A.D.) who first saw that spherical geometry must be based on great circles only. In the preceding period, either only qualitative results were reached or else graphical methods were used. One of these seems to have been based on the discovery that stereographic projection of the sphere maps circles into circles. This fact was certainly known at the beginning of our era, as is shown by the construction of mechanical clocks which represent the celestial motions in a plane, similar to the later astrolabes. Hipparchus, who had no spherical trigonometry at his disposal, may have solved spherical triangles by the method of stereographic projection.

67.While Babylonian origin is quite obvious in the arithmetical treatment of the problem of ascensions and length of daylight, a much more complex situation is encountered in the theory of the lunar motion. Insight into this part of Hellenistic astronomy is of rather recent date and far from complete. In fact, it is more appropriate to say that a new (and very promising) chapter of research has been no more than barely begun. I shall outline how this came about because it is typical of the accidental fashion in which we actually proceed, in defiance of all attempts at planning the road of research in advance. This, of course, is really not too surprising because only those objects can be reached in a systematic fashion whose outlines are already fairly well determined. And this is certainly not the case with Hellenistic astronomy and its descendants.

In 1922 Thureau-Dangin published a copy of a tablet from Uruk, now in the Louvre, dealing with the daily movement of the moon. It was discussed in 1927 by Schnabel, who pointed out that the parameters agreed with the values given by Geminus in his "Introduction" (about 100 B.C.). Several additional tablets of the same type have been identified since then, and it is possible to show that they belong to a consistent ephemeris of the moon's daily motion extending at least from the year −194/3 to −181/0. These texts are based on a zigzag function which we have already mentioned on p. 122. Its period is 248 days; in other words, it is assumed that the smallest integer number of days between,

say, two minima of the lunar velocity is 248 days. This interval
covers 9 complete oscillations of the lunar velocity, or 9 "anomalistic months", but the length of one of these anomalistic months
is not an integer number of days. Its length is given by our zigzag
function as the quotient $\frac{248}{9} = 27;33,20$ days. This value is
slightly too large and was obviously chosen in order to obtain a
conveniently round number for the difference of this zigzag
function, namely, 0;18 degrees per day. A more accurate value
can be derived from the Babylonian lunar theory itself, namely,
from the columns F and G (cf. p. 117 f.). One finds 27;33,16,26,54,
assuming full accuracy of the numbers used for F and G. Thus
it is clear that we are dealing again with two concurrent and
slightly different methods of Babylonian astronomy for the
description of the lunar motion. The one is based on highly
accurate values and is used implicitly in the lunar tables for full
and new moons. The other, for the day-by-day motion, is based
on conveniently rounded-off parameters. It is the history of this
second method which we will now analyze.

The first step was made by Schnabel in his paper of 1927.
In a short appendix he remarked that the relation 9 anomalistic
months = 248 days was not only known to Geminus but also
occurs in Hindu astronomy. This was fully in line with a discovery
which had been made by Kugler in 1900, namely, that the ratio
3:2 of longest to shortest day used by both systems in columns
C and D of the Babylonian lunar ephemerides also appears in
Hindu astronomy, though this ratio is totally incorrect for the
main parts of India.

The next step was in a new direction. E. J. Knudtzon identified
as astronomical the fragments of two Greek papyri in the Library
of the University of Lund, Sweden, and sent me photographs
shortly after the end of World War II. One of these fragments
turned out to be part of a papyrus now at the University of
California, belonging to the larger class of Demotic and Greek
papyri dealing with planetary motion (cf. above p. 95). The
other fragment, however, proved to be a new type of lunar ephemeris based on the Babylonian relation: 9 anomalistic months =
248 days. The calendar used is based on Egyptian years (of
365 days each) and, in the fragment at hand, on the regnal years

of Nero and Vespasian. The papyrus gives dates, 248 days apart;
and corresponding longitudes, 27;43,24,56° apart. The following
example of 3 consecutive lines will illustrate the procedure:

year 6 month VII day 2   ♍ 5;3,21,31
     7   III   5   ♎ 2;46,46,27
     7   XI   13   ♏ 0;30,11,23.

From this the mean lunar velocity can easily be found. Because
248 days contain 9 complete revolutions, the moon moved not
only 27;43,... during this period but 9 times 360° more. Hence
we add 54,0° to the previous number. Dividing the total 54,27;43,
24,56° by 248 gives for the daily motion 13;10,32,16,... which
is slightly less than the standard Babylonian mean value of
13;10,35° per day. Obviously the deviation noted is merely a
consequence of the fact that the period of 248 days is itself only
a rounded-off value. This could be confirmed from the very same
text. The process described so far is repeated only 11 times. After
every 11 of these ordinary steps, henceforth called D, one "big"
step Δ of 303 days was inserted, with a corresponding motion of
the moon of 11 times 360 plus 32;33,44,51°. The lunar velocity
resulting from one such big step is 13;10,36,23,..., slightly larger
than expected. This shows that we are dealing with a process of
successive approximations. Indeed, if we consider the 11 ordinary
steps plus one big step as one higher unit $C = 11D + \Delta$ of
3031 days, then we find for it the mean motion 13;10,34,51,57,...
This value is not only very close to the value 13;10,35 but we
know that 13;10,35 itself must be the result of a small rounding
off. Ptolemy's value, e. g., is 13;10,34,58,33,30,30. Similarly the
value for the anomalistic month represented by C is a much
better approximation than in D; one finds 27;33,16,21,... which
is very close to the expected 27;33,16,26,.... Thus we see that
higher groups were designed in order to avoid the accumulation
of error which was allowed in the single steps.

It can be shown that the moments chosen were the moments of
minimum velocity. Dates and positions being known for the
apogees of the moon, the positions of the moon for any other
date can be found simply by operating with the well known zigzag
function for the lunar velocity, starting with the minimum and
going linearly up and down until the given date is reached. This

computation will never lead to large errors because one always starts from the nearest minimum, whose position is well established by the general process.

Thus we see before us in a Greek papyrus a method of purely arithmetical character, based on Babylonian parameters and Babylonian schemes but adjusted to the Egyptian calendar. It is difficult to say whether the formation of groups of the type $C = 11 D + \varDelta$ is also of Babylonian origin or a later invention of the Alexandrian astronomers. Against Babylonian origin might be held the fact that the process D alone was used for more than 13 years in the preserved texts (p. 161); but this does not exclude the possibility of the existence of the improved procedure in other texts. Admission of our ignorance seems the best procedure.

How little we really know about this form of Hellenistic astronomy became evident shortly after the publication of the Lund papyrus. An investigation of dates of solstices in the Hellenistic period led me to check a Greek papyrus of the John Rylands Library, published in 1911, because it contained such data at the end. The solstices turned out to be uninteresting but it now appeared at once obvious that the main text contained, among others, the rules for the computation of the scheme of the Lund papyrus. This discovery made at least one thing clear: there existed "linear methods" of far wider extent than one could possibly have deduced from the silence of Ptolemy and his commentators.

The Ryland papyrus revealed further details of the whole method and raised new problems. It showed that the whole process, as preserved in the Greek papyri, is based on the Era of Augustus and that also the cycle of 25 Egyptian years was incorporated—the very same cycle which is known to us from the Demotic papyrus Carlsberg No. 9 (cf. p. 95). Thus the mixture of methods became still more evident for Hellenistic astronomy. But there appeared also difficulties in details; for instance, the step $\varDelta$ should have been inserted one line later, following the rules of the Ryland papyrus, than it actually appears in the Lund papyrus. We can express this also in the form that somewhere one unexpected step D was inserted.

Today I am still unable to explain every detail in these Greek texts but the general direction in which to go now appears clearer. Again returning to the study of the solar theory, I intended to investigate the transmission of Ptolemy's theory of precession to the Arabs, and it was only natural to include here the Hindu sources. This led me to the Pañca Siddhāntikā of Varāha Mihira, written about 550 A.D. We shall come back to Hindu astronomy presently (p. 173); for the moment it suffices to say that the Pañca Siddhāntikā contains certain rules for the computation of the lunar motion based on the processes now know to us from Greek sources. Thibaut, who edited the Pañca Siddhāntikā in 1889, found it very difficult to understand these passages. He eventually found the key to the problem in the book Kāla San-kalita by J.Warren. The latter had traveled extensively in Southern India and had recorded the astronomical lore of the natives in the book mentioned, published in Madras in 1825. In this book, he describes a method followed by the Tamil inhabitants of the Coromandel coast for the computation of the lunar motion. His informants no longer had any idea about the reasons for the single steps which they performed according to their rules. The numbers themselves were not written down but were represented by groups of shells placed on the ground. Thus

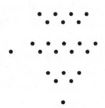

means 7 zodiacal signs and 19;5,1°. Nevertheless they carried out long computations for the determination of the magnitude, duration, beginning and end of an eclipse with numbers which run into the billions in their integral part and with several sex-agesimal places for their fractions. Simultaneously they used memorized tables for the daily motion of the sun and moon involving many thousands of numbers. Certain elements can be dated astronomically as referring to an epoch of 1200 A.D. But the Pañca Siddhāntikā already demonstrates the existence of these methods seven centuries earlier. And, finally, they go back to the Greek papyri, though the Indian sources go slightly beyond the steps known from the Hellenistic sources. One begins with the "Devaram" period $D = 248$ days. From this one forms the Calanilam period $C = 11D + \varDelta = 3031$ days. The next step is new; it consists of forming the Rasa Gherica period $R = 4C + D = 12372$ days. The value of R is almost identical with the Babylonian value for the anomalistic month; one finds 27;33,16,26, 11,.... where only the last figure is different from the expected value (cf. p. 162).

Whatever remains to be clarified, it is evident that the methods found by Warren still in existence in the 19th century are the last witness of procedures which go back through the medium of Hellenistic astronomy to Babylonian methods of the Seleucid period. I do not doubt that this specific case of the lunar theory is only one of many similar instances where very close contact between Hindu astronomy and originally Babylonian methods can be established. We shall return to this question at the end of this chapter.

When the question of contacts is raised, it might seem tempting to assume a direct relationship between India and Mesopotamia without the Hellenistic intermediary. At the present rudimentary stage of our knowledge of such questions, any definite answer is more a matter of guess and of taste than of real evidence. Nevertheless, it seems to me more plausible to assume the way through the Greek and Persian civilization of the Sasanian period than through a direct contact.

For this I may give three major arguments. First, the fact that the terminology as well as the methods of Hindu astrology are clearly of Greek origin; for example, the names of the zodiacal signs are Greek loan-words. Similarly, the basic concepts of the planetary theory of the Surya Sidhanta are influenced by the Greek epicyclic models and not by the Babylonian linear methods. This argument no longer holds for the linear methods themselves. But here—and this is my second argument—may be mentioned that precisely the coastal region from which our information about Tamil astronomy comes was a center of Roman trade. We have ample evidence for this, e. g., the anonymous "Periplus of the Erythrean Sea", written in the first century A.D., which contains a detailed account of the commerce between Egypt and India, the harbors and the kind of goods that were traded, etc. This is fully corroborated by archaeological evidence, most drastically in 1946 by the discovery of a large Roman emporium in Arikamedu in the outskirts of Pondicherry, the very same place where LeGentil learned in 1769 for the first time about the linear methods from his Tamil informants. This contact with the West has its climax in the time of Augustus and in the first century A.D., but Roman coin hoards reach into the fourth century. All this is confirmed by repeated references to the "Yavanas" (i. e., "Ionians" for "Greeks"[1])) in Hindu astronomy and Tamil literature. And finally, the chronology of Hindu astronomy: linear methods as well as trigonometric models point to the early centuries A.D., not B.C.

Whatever later discoveries might reveal, at present it seems reasonable to assume that Babylonian methods, parameters and concepts reached India in two ways, either via Persia or via the Roman sea routes, but only through the medium of Hellenistic astronomy and astrology.

68. We have already remarked that essential parameters ascribed by Ptolemy to Hipparchus are identical with the corresponding parameters of the Babylonian theory. As long as Ptolemy was our only source, it was natural to assume that Hipparchus's theory was strictly geometrical and exactly of the type later followed by Ptolemy. Now, with the linear methods coming gradually to light, the situation becomes much more complex. There can be no doubt that Hipparchus used geometrical methods; we know this, e. g., from his determination of the eccentricity of the solar orbit, as related by Ptolemy. But one can no longer exclude the possibility that he also worked with linear methods, as would be no more than natural for anyone who had access to Babylonian material. Greek astrological literature contains several references to linear methods which are associated with the name of Hipparchus. It has been customary to discard such references as apocryphal simply because one thought the great astronomer unworthy of such

---

1) This includes, of course, the Romans; also the Periplus speaks only about "Greek" ships sailing for India. Conversely, the "Romans" of Islam are the Byzantine Greeks.

elementary arithmetical schemes. I do not want to say that we should now revert to the opposite extreme and consider such references as genuine without further investigation. But it is clear that Ptolemy did not intend to be historically complete in his references to methods used by his predecessors, and that only a systematic collection of all references to Hipparchus, by Ptolemy and by all other sources, can help us to obtain a more complete picture of the contents of his writings.

69. One of the main reasons for the transmission of astronomical knowledge from one nation to another was undoubtedly the spread of the belief in astrology as the one science which gave insight into the causes of the events on earth. It has often been said that astronomy originated from astrology. I see no evidence for this theory. It seems to me much more plausible to assume that one major incentive for the development of astronomy consisted in attempts to achieve regularity in the intercalations of the lunar calendars. The best description of the true situation might be the statement that we know equally little about the origin of astrology or astronomy and that the relative influence of these two disciplines upon one another is largely a matter of conjecture. As we have explained in Chapter V, we are on safe ground for astronomy only for the Seleucid period in Mesopotamia. Almost all documents concerning astrology in Mesopotamia belong to the same period, but their number is very small compared to the astronomical texts. We have about ten horoscopes from cuneiform tablets, and still fewer texts concern astrological doctrines as they are known to us in such enormous amounts from Greek sources. In Egypt the earliest horoscopes, Demotic or Greek, are from the reign of Augustus. To the same period belong the earliest astrological treatises and a general discussion about the validity of astrological doctrine. The rapid spread and enormous development of astrology during the first Roman imperial period is paralleled in the spread of Christianity, Mithraism, and related creeds. In neither case can the speed and extent of this spread be utilized as a chronological argument.

The only chronological criteria can come from the texts themselves. One of these criteria is the arrangement of the planets. In the cuneiform texts of the Seleucid period the standard arrangement is

Jupiter—Venus—Mercury—Saturn—Mars.

The reason for this arrangement is unknown; the commonly given explanation that the first two planets are beneficiary, the last two malevolent, with Mercury doubtful, does not appear in cuneiform sources. The ordinary arrangement in the Greek horoscopes is

Sun—Moon—Saturn—Jupiter—Mars—Venus—Mercury,

except for cases where an arrangement is chosen which depends on the special horoscope, that is, following the positions of the celestial bodies in the zodiac at the given moment. I think these two lists reflect the difference between the two astronomical systems very clearly. The Babylonian system has nothing to do with the arrangement in space. The Greek system, however, obviously follows the model which arranges the planets in depth according to their periods of sidereal rotation. This is reflected even in the arrangement of the days of the planetary week which we still use today. Here the Sun is placed between Mars and Venus, and the Moon below Mercury. Every one of the 24 hours of a day is given a "ruler" following this sequence. Beginning, e. g., with the Sun for the first hour one obtains

day 1	hour 1	2	3	4	5 ...	24
	Sun	Venus	Mercury	Moon	Saturn...	Mercury
day 2	hour 1	2	3	....		24
	Moon	Saturn	Jupiter	....		Jupiter
day 3	hour 1	.... etc.				
	Mars					

The "ruler" of the first hour it then considered to be the ruler of the day and thus one obtains for seven consecutive days the following rulers

Sun Moon Mars Mercury Jupiter Venus Saturn

which is our sequence of the days of the week and also the arrangement of the planets in Hindu astronomy.

Here we have a system which is obviously Greek in origin not only because it is based on the arrangement of the celestial

bodies according to their distance from the earth but also because
it supposes a division of the day into 24 hours, a form of reckoning
which is not Babylonian but a Hellenistic product of ultimately
Egyptian origin (cf. p. 86). It is totally misleading when this
order is called "Chaldean" in modern literature.

As we have said before, the astrology which is known to us
from the Assyrian period is quite different from the Hellenistic
personal astrology. The predictions concern the king and the
country as a whole and are based on observed astronomical
appearances, not on computation and not on the moment of
birth. In addition, the zodiac never appears. Hellenistic horoscopes, however, concern a specific person and depend upon
the computed position of the seven celestial bodies and of
the zodiacal signs in their relation to the given horizon, for
a given moment, the moment of birth. Around this is woven
an enormous system of doctrines concerning the evaluation of
these data and of secondary data which can be derived by
all kinds of artifices so as to obtain a greater variety of possibilities. It is interesting to observe that the actually preserved
horoscopes contain very little, if anything at all, of these theoretical
speculations. The great majority contain nothing but the bare
results of the computations for the given moment. This makes
these texts useful documents for the study of purely astronomical
and chronological questions, but they help us very little for the
history of astrology as such and of the astronomical methods
imbedded in its doctrines. Nevertheless, the patient work of
scholars like Bouché-Leclercq, Cumont, Boll, Bezold, Kroll, Rehm
and many others has shown the existence of different components
of diverse origin. There exist predictions which fit only very
specific circumstances, like the destruction of the Persian empire
by Alexander or the wars between his successors in Syria;
finally, there is a great mass of references to Egypt under the
rule of the Ptolemies. The references to constellations, especially
their simultaneous risings and settings, made it possible to distinguish between two widely different celestial maps, a "sphaera
barbarica" and a "sphaera graecanica". Yet the fact remains that
the evidence for direct borrowings from Babylonian concepts
remains exceedingly slim. The main structure of the astrological
theory is undoubtedly Hellenistic. On p. 68 we have discussed

the remnants of the oldest available catalogue of stars, contained
in the astrological writings which go under the name of Hermes
Trismegistos. The fact that these star coordinates correspond to
the time of Hipparchus or his direct followers (cf. p. 69) is an
added argument for the origin of this major work of astrology
in the second century B.C.

Though it is quite plausible that the original impetus for
horoscopic astrology came from Babylonia as a new development from the old celestial omens, it seems to me that its actual
development must be considered as an important component of
Hellenistic science. To a modern scientist, an ancient astrological
treatise appears as mere nonsense. But we should not forget that
we must evaluate such doctrines against the contemporary background. To Greek philosophers and astronomers, the universe
was a well defined structure of directly related bodies. The concept
of predictable influence between these bodies is in principle not
at all different from any modern mechanistic theory And it stands
in sharpest contrast to the ideas of either arbitrary rulership of
deities or of the possibility of influencing events by magical
operations. Compared with the background of religion, magic and
mysticism, the fundamental doctrines of astrology are pure
science. Of course, the boundaries between rational science and
loose speculation were rapidly obliterated and astrological lore
did not stem—but rather promoted—superstition and magical
practices. The ease of such a transformation from science to
humbug is not difficult to exemplify in our modern world.

70. To the historian of civilization, astrology is not only one
of the significant phenomena of the Hellenistic world but an
exceedingly helpful tool for the investigation of the transmission
of Hellenistic thought. As an example may be quoted Abū
Ma'shar, who died in 886 and is an early representative of
Hellenistic astrology among the Arabs. Boll has shown that he
utilized a Persian translation, made in 542, of the "Sphaera
Barbarica" of Teukros. Thus Abū Ma'shar becomes an important source for an early Hellenistic lore of constellations. The
famous astrological paintings in the Palazzo Schifanoja in Ferrara,
made in the second half of the 15th century, are influenced by
the doctrines of Abū Ma'shar's astrology. On the other hand,
his writings were translated into Latin, into Greek, into Hebrew,

and from Hebrew into Latin. These "translations" are often only
freely handled versions, incorporated in other material of diverse
origin. There exist even complete cycles of translations and
borrowings from Greek back into Greek. For example, chapters
from an astrological poem in hexameters by Dorotheos of Sidon
(first century A.D.) were used by Abū Ma'shar, who in turn
provided the prototype for a Byzantine dialogue "Hermippos".
Similar cycles can be established for astronomical tables and
treatises which reached Byzantium in the 13th century.

There is found, however, in Abū Ma'shar's writings another
component which makes them of great interest for our problem
of tracing the transmission of Hellenistic science. Indian asterisms
appear in Abū Ma'shar, and their source is found in the astrological writings of Varāha Mihira, the same author of the sixth
century A.D. in whose astronomical work we found the use of
the linear methods for the lunar motion, otherwise known to
us from Greek papyri and finally from cuneiform tablets. Following the unmistakable traces of very specific astrological doctrines, one can reconstruct the road which connected Hellenistic
Mesopotamia with Hellenistic Egypt, with pre-Islamic Persia,
and with India. We are obviously entitled to assume that the
same road was followed by the transmission of mathematical
astronomy even if no more is available to us than the two extremal
ends in Mesopotamia and India.

In the case of the lunar theory, at least one missing link, the
papyri, were available. In the case of the planetary theory,
however, not even that much is known from Greek sources.
Nevertheless, we can now understand whole sections in Varāha
Mihira's Pañca Siddhāntikā by means of the Babylonian planetary texts. We have seen how the planetary phenomena were
described in Babylonian texts by means of step functions which
we generally called "System A". Precisely the same idea is
found in the Pañca Siddhāntikā. The same holds for fundamental
period relations and even for special parameters. A few examples
may be quoted. For Saturn and Jupiter we know from cuneiform
sources the synodic periods

$$\frac{4,16}{9} = 28;26,40 \quad \text{and} \quad \frac{6,31}{36} = 10;51,40$$

and for Venus the synodic arc of 3,35;30°. All three values are used by Varāha Mihira.

Very strange values seemed to be assigned by Varāha Mihira to the duration of the synodic revolutions of the planets; for example,

$$
\begin{array}{ll}
\text{Mars} & 768\tfrac{3}{4} \ \text{days} \\
\text{Mercury} & 114\tfrac{6}{29} \ \text{days} \\
\text{Jupiter} & 393\tfrac{1}{7} \ \text{days} \\
\text{Venus} & 575\tfrac{1}{2} \ \text{days} \\
\text{Saturn} & 372\tfrac{2}{3} \ \text{days.}
\end{array}
$$

The comparison of these numbers with Babylonian parameters immediately gives the solution of the problem. Not "days" are meant, but degrees. Indeed the mean synodic arc for Mercury is $1,54;12,24,\ldots = \dfrac{55,12}{29}$ which is exactly the Hindu value; and the same agreement is found for Venus. The mean synodic arc for Mars, however, is, according to the Babylonian theory, $6,48;43,18,\ldots$ which is very close to $6,48;45 = 408\tfrac{3}{4}$ whereas the Pañca Siddhāntikā gives $768\tfrac{3}{4}$. But the difference is $360°$, or one complete rotation. This explains also the replacement of "degrees" by "days"; the Hindu "days" are simply sexagesimal fractions of a sidereal year, or, expressed differently, the mean solar motion is assumed to be $1°$ per day. This is confirmed for the case of Jupiter and Saturn. Instead of $393\tfrac{1}{7}$ days we consider only $33\tfrac{1}{7} = 33;8,34,\ldots°$ and this is again in good agreement with the Babylonian value $33;8,45°$ for the mean synodic arc. Similarly for Saturn: $372\tfrac{2}{3} - 360 = 12;40°$ as compared with $12;39,22,30°$ in the Babylonian theory.

We stand today only at the beginning of a systematic investigation of the relations between Hindu and Babylonian astronomy, an investigation which is obviously bound to give us a greatly deepened insight into the origin of both fields.

71. The fact that a close relationship between Babylonian linear methods and sections of the Pañca Siddhāntikā can be established is only one facet of the general problem of the evaluation of the role of Hindu astronomy in the history of science. In it the Pañca Siddhāntikā of Varāha Mihira is of special importance because it forms a chronological fixed point of the first importance. Varāha Mihira's date is well established by his use of the year 427 Saka Era = 505 A.D. as an epoch and by considerations which lead to about 590 A.D. as an upper limit for his lifetime. But it is important not only that the Pañca Siddhāntikā is in this way a well dated and comparatively early document; it is also a historical source of a unique character in Hindu astronomy. Its name indicates that it is based on five Siddhāntas, and it actually contains a summary of the contents of the five great astronomical treatises which were in existence in Varāha Mihira's time. Thus we have here an early historical report on source material which is no longer extant, or at least no longer extant in exactly the same form. On the other hand Varāha Mihira is also one of the main sources of al-Bīrūnī's report on Hindu astronomy and astrology, written about 1030 A.D. Consequently Varāha Mihira occupies a central role for the study of Hindu astronomy.

The Sūrya Siddhānta has to be considered as the main canon of Hindu astronomy. It is supposed to have been revealed by the Sun (Sūrya) at the end of the Golden Age (2163102 B.C.) to a Maya Asura. Some manuscripts contain the additional command of the Sun to Maya: "Go therefore to Romaka-city, your own residence; there, undergoing incarnation as a barbarian, owing to a curse of Brahma, I will impart to you this science"[1]). This is closely paralleled by a passage in Varāha Mihira: "The Greeks, indeed, are foreigners, but with them this science [astronomy] is in a flourishing state"[2]). The origin of the Sūrya Siddhānta is dated by modern scholars to about 400 A.D. while its present version may be as late as about 1000 A.D. That this work contains several much older and very primitive sections which it combines rather startlingly with the Greek theory of epicyclic motion has been apparent to all scholars since al-Bīrūnī, who characterizes Hindu mathematical and astronomical literature somewhat drastically as "a mixture of pearl shells and sour dates, or of pearls and dung, or of costly crystal and common pebbles"[3]).

[1]) Sūrya Siddhānta I, 6 (Burgess).
[2]) Bṛhat Saṁhita II, 15 (Kern). An insignificantly different translation is proposed in Isis 14 (1930) p. 391 or in the translation by V. Subrahmanya Sastri, Bangalore 1947, p. 19 (Sloka 14).
[3]) India I (Sachau I, p. 25).

Though the Greek influence on the Sūrya Siddhānta is evident, it is equally obvious that the Greek theory has undergone a quite independent transformation in many details both with respect to the values of numerical constants and to the general theory. That modifications of this type went on continuously is explicitly proved by the comparison of the existing Sūrya Siddhānta with the summary of Varāha Mihira in the Pañca Siddhāntikā. On the whole, however, it seems as if the Sūrya Siddhānta were the most consistently modified Hindu treatise. What we know about the Romaka and the Paulisa Siddhānta from the material which was incorporated in the Pañca Siddhāntikā seems to bring these treatises closer to the Hellenistic sources. Their names support this view; the "Romans" are, of course, the Greeks of the Roman or Byzantine empire; and al-Bīrūnī considers the Paulisa Siddhānta the work of Paulus Alexandrinus, an astrologer of the fourth century A.D. Thus we obtain again as the period of contact roughly the time of origin of the Sūrya Siddhānta, i. e., about 400 A.D. And it is perhaps significant that the earliest occurrence of the place value notation can be traced back again to the Paulisa Siddhānta.

On the other hand, it has long been recognized that the borrowings of Hindu astronomy from Greek astronomy show no influence of the refinements of the Ptolemaic theory. Also the astrological theory reflects, at least partially, the oldest strata of Hellenistic doctrine. This would lead to a period between Ptolemy (150 A.D.) and Hipparchus (150 B.C.) or even slightly earlier. This seems to leave a serious gap of several centuries between the date of the Hellenistic sources and reception by the Hindus. There is, however, increasing evidence forthcoming to bridge this gap. Ibn Yūnus, the great Arabic astronomer (died 1009), famous as author of the Hakemite tables, quotes Persian observations of the apogee of the solar orbit made about 470 A.D. and 630 A.D. Nallino has shown that both Teukros and Vettius Valens were translated into Pehlevi, the pre-Islamic or Middle-Persian Iranian language. We have already seen that Teukros (first century B.C.?) accounts for early elements in Greek astrology; Vettius Valens is a younger contemporary of Ptolemy but uninfluenced either by the Almagest or by the Tetrabiblos. The oftquoted passage where Vettius Valens declares that he "did not compute eclipse tables himself but used Hipparchus for the sun,

Sudines and Kidenas for the moon", relates him directly with the linear methods of the Babylonians, and with whatever geometrical or arithmetical methods were used by Hipparchus. If we assume that these sources reached Persia without being modified by the scientific theories of Ptolemy, then we have a satisfactory explanation for the main features both of the linear and of the geometrical methods found in the Pañca Siddhāntikā.

72. The way back leads again via Persia. It is well known that the scientific activity of Islam originated under the Abbasid Khalifate in Baghdad; al-Khwārizmī, Thābit ibn Qorra, Abū Ma'shar belong to this period (9th century).

Al-Khwārizmī's astronomical tables have been preserved through Latin translations; they show a curious mixture of Hindu and Greek methods. The relation between his mathematical writings and the Hellenistic tradition has already been mentioned. A century later appears al-Bīrūnī, another native of Khorazm. He not only transmitted Hindu knowledge to the West, but the tells us that "most of their books are composed in Sloka [verses], in which I am now exercising myself, being occupied in composing for the Hindus a translation of the books of Euclid and of the Almagest, and dictating to them a treatise on the construction of the astrolabe, being simply guided herein by the desire of spreading science". On the other hand, al-Bīrūnī translated an astrological work of Varāha Mihira into Arabic[1]).

The history of the transmission of Hellenistic science throughout the Islamic world need not be told here. The general trend is no longer in doubt and has often been described. What is less generally known, however, is the fact that for every specific question of astronomical or mathematical theory we are still groping in the dark because of a most deplorable lack of edited source material. With the splendid exception of al-Battānī's tables, none of the great astronomical tables of the Middle Ages— Arabic or Latin, Hebrew of Greek—is available in modern editions for the period between Ptolemy and Copernicus. The history of the ancient mathematical sciences is a field in which one need not go far to find fertile soil ready to be cultivated.

73. We have come to the end of our discussion, which has brought us back to the civilization of the Middle Ages from which

[1]) India XIII and XIV (Sachau I, p. 137 and p. 158).

our journey started. By patiently following the connections of mathematical and astronomical theory we moved from period to period and from civilization to civilization. Our road often went parallel to the road pointed out by historians of art, religion, alchemy, and many other fields. This is not surprising. It only underlines the intrinsic unity of human culture. The role of astronomy is perhaps unique only in so far as it carried in its slow but steady progress the roots for the most decisive development in human history, the creation of the modern exact sciences. To follow this specific aspect of cultural history seems to me worthy of our efforts, however fragmentary our results may be.

In the "Cloisters" of the Metropolitan Museum in New York there hangs a magnificent tapestry which tells the tale of the Unicorn. At the end we see the miraculous animal captured, gracefully resigned to his fate, standing in an enclosure surrounded by a neat little fence. This picture may serve as a simile for what we have attempted here. We have artfully erected from small bits of evidence the fence inside which we hope to have enclosed what may appear as a possible, living creature. Reality, however, may be vastly different from the product of our imagination; perhaps it is vain to hope for anything more than a picture which is pleasing to the constructive mind when we try to restore the past.

BIBLIOGRAPHY TO CHAPTER VI

A good introduction to the political and cultural history of Hellenism is W. W. Tarn and G. T. Griffith, Hellenistic Civilisation, London, Arnold, 1952.

The following work gives an excellent picture of the multiple paths of scientific contacts during the Middle Ages in the West: Charles Homer Haskins, Studies in the History of Mediaeval Science. 2nd ed. Cambridge, Harvard University Press. 1927.

One of the most remarkable books about Hindu science and its relations to the Muslims is Alberuni's India, An Account of the Religion, Philosophy, Literature, Geography, Chronology, Astronomy, Customs, Laws, and Astrology of India about A.D. 1030; an English edition with notes and indices by Edward C. Sachau, London 1910. For special works on Hindu astronomy cf. the notes ad No. 67. Cf. also M. Steinschneider, Zur Geschichte der Übersetzungen aus dem Indischen ins Arabische etc., Zeitschr. d. Deutschen Morgenländischen Gesellschaft 24 and 25 (1870 and 1871).

For the influence of Iranian thought during the Hellenistic period see Joseph Bidez et Franz Cumont, Les mages hellénisés; Zoroastre, Ostanès et Hystaspe d'après la tradition grecque; 2 vols. Paris, 1938.

For the early relations between Greeks and India see W. W. Tarn, The Greeks in Bactria and India, 2nd. ed. Cambridge 1951 and A. K. Narain, The Indo-Greeks, Oxford 1957.

As a summary of Hindu science should be quoted the chapter on science by W. E. Clark in "The Legacy of India" (edited by G. T. Garratt, Oxford 1937). A detailed summary of the literature up to 1899 is given by G. Thibaut in his article "Astronomie, Astrologie und Mathematik" in vol. III, 9 of the "Grundriss der Indo-Arischen Philologie und Altertumskunde". Very useful is James Burgess, Notes on Hindu Astronomy and the History of our Knowledge of It (J. of the Royal Asiatic Soc. of Great Britain and Ireland, 1893, p. 717–761) where one finds complete references to the early literature which contains much important information which is no longer available otherwise.

The translation by E. Burgess of the Sūrya Siddhānta, quoted below p. 186, contains extensive commentaries which must be read by any serious student of this subject. For the "linear methods" in Hindu astronomy cf. the references to Le Gentil and Warren on p. 186. For the form which the Greek theory of epicyclic motion of the planets took in India and then in al-Khwārizmī, see O. Neugebauer, The transmission of planetary theories in ancient and medieval astronomy, Scripta Mathematica, New York, 1956.

E. S. Kennedy, A survey of Islamic astronomical tables, Trans. Amer. Philos. Soc., N.S. 46 (1956) p. 123–177 is a publication which shows the great wealth of material still available but barely utilized for the investigation of medieval astronomy, its Greek, Islamic and Hindu sources and their interaction.

NOTES AND REFERENCES TO CHAPTER VI

ad 61. The date of Heron Alexandrinus is fixed in the second half of the first century A.D. by the fact that he quotes as an example the lunar eclipse of A.D. 62 March 13 [cf. Neugebauer, Kgl. Danske Vidensk. Selsk., hist.-filol. medd. 26, 2 and 7 (1938 and 1939) and A. G. Drachmann, Centaurus I (1950) p. 117–131].

Our only certain knowledge about the date of Diophantus rests upon the fact that he quotes Hypsicles (opera I p. 470, 27 and 472, 20, ed. Tannery) and that he is quoted by Theon Alexandrinus (Comm. in Almag. I, p. 453, ed. Rome). The date of Hypsicles can be roughly estimated as about 150 B.C.; Theon's date is fixed by the solar eclipse of 364 June 16. One now usually follows Tannery, who argued for a date about 250 A.D. Tannery's argument is based on an admittedly corrupt passage from Psellus (about 1050) and is in general very hypothetical, as has been pointed out by J. Klein in Quellen und Studien z. Gesch. d. Math., ser. B vol. 2 (1934) p. 133 note 23. Klein himself favors an earlier date and suggests possible contemporaneity with Heron. Klein's arguments are largely based on the stylistic similarities between Heron and Diophantus, an argument which has lost much of its weight since we have become aware that both authors more or less represent common Oriental-Hellenistic tradition.

The dedication in both cases to a certain Dionysius is not decisive, not only because this name is very common but also because of the use of different titles given to Dionysius.

A medieval dating is preserved in the "History of Dynasties" of Bar Hebraeus (1226–1286), also called Abū 'l-Faraj, a learned Jacobite bishop. His "History of Dynasties" is written in Arabic[1]) and is an abridged and modified version of his Syriac "Chronography"[2]). Only the Arabic version quotes Diophantus and his "Algebra", referring it to the time of Julian the Apostate (361–363). This date is just barely reconcilable with the upper limit for Diophantus's lifetime, mentioned above. That Bar Hebraeus was a competent mathematician and astronomer is well known[3]), but we do not know on what authority his chronological statement was based. I see no possibility for confirming or disproving it and it seems to me that we must admit that Diophantus cannot be dated with any accuracy within 500 years.

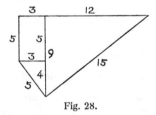

Fig. 28.

Writings of the type of Heron's "Geometry" were undoubtedly widespread in antiquity and formed the backbone of instruction in elementary mathematics. This explains the relatively large number of papyrus fragments containing such texts. As an example can be shown an unpublished papyrus of the Cornell Collection (cf. Pl. 12). The figure in the lower part of the right column is a typical example of the building up of a more complicated example from the simplest cases (cf. Fig. 28).

It is precisely from the construction of such examples that we can demonstrate direct relations between geometrical treatises. For example, the concept of isosceles triangle is illustrated by a triangle of side 10 in Heron, Metrica I, XVII (opera III p. 48, 49); in Heron, Geometrica 10 (opera IV, p. 224, 225); al-Khwārizmī, Algebra (Rosen p. 80); al-Bīrūnī, Astrology 22 (Wright). The general triangle of sides 13, 14, and 15 is used by Heron, Metrica I, V and VIII; Geometrica 12; Mishnat ha-Middot 9 (Gandz p. 46); Mahāvīra VII, 53 (ed. Rangacharya p. 199); al-Khwārizmī p. 82; Bhascara, Līlāvatī VI, 165, 168 (Colebrooke p. 71, 73). In fact this triangle is composed of two Pythagorean triangles of sides 13, 5, 12, and 15, 9, 12 respectively. They occur again in two problems of the Pap. Ayer (Am. J. of Philology 19, 1898, p. 25 ff.).

[1]) Edited with Latin translation by Edward Pococke, Oxford 1663.
[2]) Edited with English translation by A. W. Budge, Oxford 1932.
[3]) Cf., e. g., F. Nau, Le livre de l'ascension de l'esprit ... Cours d'astronomie rédigé en 1279 par ... Bar Hebraeus (Bibliothèque de l'école des hautes études 121, Paris 1899/1900).

The fact that a Hebrew treatise is part of this tradition (cf. S. Gandz, The Mishnat ha-Middot; Quellen und Studien zur Geschichte der Math., Abt. A, vol. 2, 1932) is perhaps not as isolated a phenomenon as it may appear. The Heronic corpus itself contains several references to Hebrew units of measure (Heron, Opera V p. 210–219). Similarly, many concepts of Judaism are found in the magical papyri and in related practices involving numbers and the alphabet, the so-called gematria. There is certainly some basis in the ancient terminology which uses "mathematici" in the sense of magicians or astrologers.

Another example for the transmission of mathematical knowledge is found in al-Bīrūnī's India XV (Sachau I p. 168 f.), where he says "Brahmagupta relates with regard to Āryabhata . . . that he fixed the circumference as 3393, . . . the diameter as 1080". The reason for this expression, which contains a common factor 9, becomes obvious if one recalls that 1080 is an important metrological unit in oriental astronomy. As an example can be mentioned the division of one hour in 1080 parts (chelakim) in Hebrew astronomy. The sexagesimal equivalent of $\frac{3393}{1080}$ is 3;8,30 which is the approximation of π used in the Almagest (VI, 7).

In the same section al-Bīrūnī states that "Pulisa employs . . . in his calculations . . . 3 $\frac{177}{1250}$. . . . The same relation is derived from the old theory, which Ya'kub ibn Tārik mentions in his book, Compositio Sphaerarum, on the authority of his Hindu informant . . ." Indeed the same value is used by al-Khwārizmī (Algebra, ed. Rosen p. 72 + 198 f.). Its decimal equivalent is 3.1416, its sexagesimal equivalent 3;8,29,45,36. For a discussion of these and related values see B. Datta, Journal and Proceedings of the Asiatic Society of Bengal, N.S. 22 (1926) p. 25–42, esp. p. 26 f.

The use of the value 3 for π is commonly found in Hellenistic texts and goes back to Old Babylonian mathematical texts. Another example common to Babylonian and Heronic mathematical education is the computation of the volume of "ships" of prismatic form (cf. Neugebauer, MKT II p. 52 and Heron, Opera V pp. 56, 128, 130, 172).

Traditionally it is assumed that Hellenistic science reached the Arabs through the intermediate stage of Syriac versions of the Greek works. Though this may be so in many cases it is certainly an oversimplification. G. Bergsträsser edited and translated in 1925 a work of Hunain ibn Ishāq concerning the translations of the writings of the physician Galen, a contemporary of Ptolemy (Abh. für die Kunde des Morgenlandes 17, No. 2; cf. also Max Meyerhof, New Light on Hunain ibn Ishâq and his Period, Isis 8, 1926, p. 685–724). Hunain, a Nestorian, played a central role in the early phase of Arabic translation. He was born in 809 and died in 877; his search for Greek manuscripts led him all over the Near East and to Alexandria. He must have seen and compared hundreds of them and accumulated a large collection of his own. From his report one learns how these translators worked, comparing defective manuscripts, restoring, explaining, excerpting. There is no such simple sequence, Greek → Syriac → Arabic, visible. By far the greatest number of works is directly translated from Greek either into Arabic or into Syriac. There are many cases where Syriac translations were the basis for Arabic versions but also the opposite order occurs. Hunain's report covers about 50 years and concerns more than 130 books ascribed to Galen or his school. Only 10 titles were not translated according to Hunain. From the rest 179 Syriac and 123 Arabic versions were listed by Hunain, of which he himself contributed 96 and 46 translations respectively, not counting revised versions. Of these translations 81 were made for Arabic customers, 73 for men who read Syriac.

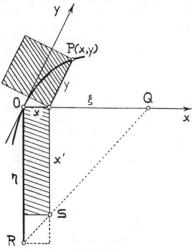

Fig. 29.

What is said here about medical literature may or may not hold for mathematical or astronomical works. One must not forget that practical need and local conditions might have been widely different for different fields of learning. For astronomy, e. g., a transmission via India undoubtedly plays a great role whereas we have very little evidence for a Syriac intermediary.

ad 62. The theory of "application of areas" attained great importance in ancient mathematics because of the discovery that the conic sections could be incorporated in this theory. Indeed, our modern names ellipse, hyperbola, and parabola are directly taken from this theory. The case of the ellipse might be quoted as an example. Assume as given two "coordinated" directions (from which our use of the word "coordinates"), here denoted as the x- and y-direction (Fig. 29). Let ξ and η be two given parameters, to be represented by line segments in the x-direction and perpendicular to it, respectively. Then a point $P(x, y)$ will be a point of an ellipse with $OQ = \xi$ as diameter and with the y-direction as conjugate direction if the area of the square with side y equals the area of the rectangle xx' which is "applied" to $OR = \eta$ in such a way that a small rectangle (RS) remains whose sides have the given ratio of η to ξ. This is only a slight generalization of the "elliptic" case of the application of areas described in the text (p. 149 f.), where the remaining rectangle was a square. In our notation P(x, y) is determined by

$$y^2 = xx' \qquad \frac{x'}{\xi - x} = \frac{\eta}{\xi}.$$

In the case of the hyperbola one requires an excess for the rectangle xx', whereas the parabola corresponds to exact equality of y^2 and $x\eta$. For details and figures cf. O. Neugebauer, Apollonius-Studien, Quellen und Studien zur Geschichte der Mathematik, Abt. B vol. 2 (1932) p. 215–254.

The historical sequence of these discoveries seems to be as follows. Since Old-Babylonian times the knowledge of solving the main types of quadratic equations existed. The discovery of irrational quantities led to the geometrization of these methods in the form of application of areas (4th century B.C.). Shortly afterwards the conic sections were discovered, as I think, from the investigation of sun dials (cf. p. 226). At any rate the conic sections were at first considered as curves in space, unrelated to algebraic problems. Finally the relation to the application of areas was established, as found in Apollonius (3rd century B.C.).

Figures which illustrate the configurations in space from which the relations between the plane areas were derived are given in Quellen und Studien zur Geschichte der Mathematik, Abt. B vol. 2 (1932) p. 220 f.

ad 63. The relationship between mathematics and Plato's theory of Ideas has been the subject of innumerable publications. For a realistic discussion of the whole problem cf. H. F. Cherniss, The Riddle of the Early Academy, University of California Press, 1945.

The uneasiness which a good Platonist felt when he was dealing with astronomical theories based on observations is nicely seen in the introduction of Proclus to his "Hypotyposis" (Greek with German translation by Manitius, Leipzig, Teubner, 1909).

ad 64. The eight-shaped curve on which a planet P moves according to the combined motion of two inclined spheres was called "Hippopede". Its qualities can easily be deduced from the following consideration (Fig. 29 a). We project the motion of the planet on the plane of the great circle RP_0T, which corresponds to the horizontal circle in Fig. 24 p. 154. The planet itself moves on the inclined plane RP' which appears in our projection as an ellipse. If the motion of the planet started at R, we can obtain its position after a motion of angle α by first moving by this amount from R to P' and then turning P' back by $-\alpha$. Since P' as point of an ellipse is the vertex of a right triangle $QP'P_0$, we see that P is the vertex of a congruent right triangle SPR. Thus P runs once through a circle of diameter RS while P' moves through the semicircle RT. And since the angle at Q and therefore also at S is α, we see that the arc from R to P is 2α. Thus P moves on its circle with twice the angular velocity of P'. Since P in Fig. 29 a represents only the projection of the planet, the orbit on the sphere is the intersection of the sphere with a straight circular cylinder of diameter SR. This gives the two loops of the Hippopede, with R as double point (Fig. 24).

A detailed comparison between the Ptolemaic theory of the motion of the moon and the modern theory was given by A. F. Möbius, Gesammelte Werke IV

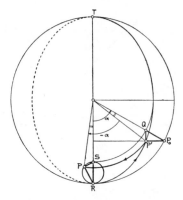

Fig. 29a.

("Mechanik des Himmels"). Cf. also Paul Kempf, Untersuchungen über die ptolemäische Theorie der Mondbewegung, Thesis, Berlin 1878; C. J. Schumacher, Untersuchungen über die ptolemäische Theorie der unteren Planeten; Münster, Aschendorff, 1917; P. Boelk, Darstellung und Prüfung der Mercurtheorie des Claudius Ptolemaeus. Thesis, Halle, 1911. The transformation from the geocentric to the heliocentric system is often hailed as one of the greatest discoveries of modern science, though foreshadowed by the Greek genius. In fact, however, the equivalence of these two modes of description of the observable phenomena had scarcely been forgotten by the astronomers of the Middle Ages. Āryabhaṭa (about 500 A.D.) argued for a movable and rotating earth (Āryabhaṭīya IV, 8; trsl. Clark p. 64 ff.) and al-Bīrūnī (1030 A.D.) remarks in his "India" (trsl. Sachau I p. 276 ff.) rather casually, "Besides, the rotation of the earth does in no way impair the value of astronomy, as all appearances of an astronomic character can quite as well be explained according to this theory as to the other."

ad 65. Cf. also O. Neugebauer, Notes on Hipparchus. Studies presented to Hetty Goldman, New York 1956, p. 292–295.

ad 66. It seems to me possible that a horoscope for the year 137 A.D. (P. Paris. 19, lines 11/12) has preserved for us the ancient name for the "linear methods". If we are correct in restoring this passage, the astrologer tells us that he computed the position of the sun according to the method of "greatest and smallest [velocity]". This would be an appropriate description of a linear zigzag function, as used for the solar velocity in System B of the Babylonian theory.

The development of the ancient and mediaeval concept of "clima" can be described quite simply. In Babylonia originated the norm that the ratio M:m of longest to shortest daylight was 3:2 as well as the two "systems" A and B for the rising times which determine the variation of length of daylight during the year.

These arithmetical methods were transplanted to Alexandria. In the second century B.C. we find Hypsicles using the System A of rising times for the computation of the length of daylight in Alexandria, the only modification being that

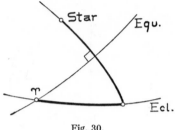

Fig. 30.

M:m is now given the value 7:5, or $M = 3,30° = 14^h$, $m = 2,30° = 10^h$. The same can be done, of course, for any given ratio M:m. But it is characteristic that this expansion to arbitrary localities was again made in a strictly linear fashion by varying M in constant steps of 4°. In this way seven zones or "climates" were distinguished, number one being Alexandria (M = 3,30), number two M = 3,34, and so forth to 3,38 ... up to a seventh climate with M = 3,54 (about 43° north). Unfortunately, the classical value M = 3,36 for Babylon does not fall into this scheme. Consequently, we find a second type of division, again in steps of 4°, but now with Babylon and M = 3,32 as the starting point (though always called "second clima", a fact which demonstrates the Alexandrian origin of all these devices). Since we have a choice, in all cases, between System A and System B for the rising times, we have in principle four possibilities for the rising times in each "clima": either M = 3,30 (Alexandria) plus a multiple of 4°, or M = 3,36 (Babylon) plus a multiple of 4° to be used either with rising times of System A or of System B. This explains the great variety of figures found in the texts for the rising times for different climates.

As in so many other cases, no trace of these primitive schemes is tolerated in the Almagest. There Ptolemy gives (in Book II) a table of rising times for ten zones (beyond the equator, which is the so-called "sphaera recta") such that M increases from zone to zone by one-half hour. Alexandria with $M = 14^h$ (here called "the low country of Egypt") is obviously part of this system, which is extended until $M = 17^h$ ($\varphi = 54;1°$). In practice, however, again only seven of these zones were accepted as "climates", this time beginning with the second of Ptolemy's zones ("Meroe" with $M = 13^h$) and ending with $M = 16^h$ or M:m = 2:1 (Southern Russia). In this system Alexandria lies in the third clima, the next being "Rhodes", the fifth "Hellespont" ($M = 15^h$).

In later periods new additions were introduced according to special needs. For example, for Byzantium $M = 15;15^h$ was adopted, but such new zones did not interfere with the standard system of the "seven climates" which remained a basic element in mediaeval geography. Only gradually did the geographical latitude replace the greatest length of daylight as the defining parameter of a locality. This development reaches its climax in the tables of rising times of al-Kāshī[1] which proceed from degree to degree of geographical latitude, from sphaera recta to $\varphi = 60°$.

[1] Unpublished; cf. E. S. Kennedy, Trans. Am. Philos. Soc. N.S. 46 p. 164.

The inventor of the system of "seven climates" is unknown: a fact which is demonstrated by the great number of theories forwarded to solve this problem. All that is reasonably certain is that it originated in Alexandria. There is no trace in Babylonian astronomy of varying values of M according to geographical latitude. The linear methods are frequently used in astrological literature and it is therefore not surprising that we find the rising times for Babylon (System A) even in India (Varāha Mihira, Brihat Jataka I, 19). It is incorrect, however, to distinguish "astrological" and "geographical" climates.

For the history of the climates see Ernst Honigmann, Die sieben Klimata (Heidelberg, Winter, 1929) and my paper "On Some Astronomical Papyri and Related Problems of Ancient Geography" in the Trans. Am. Philos. Soc., N.S. 32 (1942) p. 251–263. For the rising times in Babylonian astronomy cf. my paper in J. Cuneiform Studies 7 (1953) p. 100–103 and ACT I p. 240.

The great importance of the problem of ascensions is also visible in the only preserved writing of Hipparchus, his commentary to Aratus. In a truly masterful paper, H. Vogt has investigated Hipparchus's methods ("Versuch einer Wiederherstellung von Hipparchs Fixsternverzeichnis" Astron. Nachrichten 224, 1925, cols. 17–54). As Boll demonstrated, Hipparchus's catalogue of stars recorded about 850 stars. From this last catalogue about 350 stars are mentioned in the Aratus Commentary. Vogt investigated 879 numerical data concerning these stars; 471 of them are spherical coordinates, 408 concern simultaneous risings or settings or culminations. Hipparchus does not use the ecliptic coordinates "longitudes" and "latitudes" which have been standard in star catalogues ever since the Almagest. Of the 471 preserved numbers, 64 are declinations, 67 are right ascensions, and 340 are ecliptic arcs determined by the intersection with the ecliptic of the circle of declination through the star (cf. Fig. 30).

The ancient astronomers rightly had greater confidence in the accuracy of their mathematical theory than in their instruments. As an example may be quoted the fact that Ptolemy determines the ecliptic coordinates of fixed stars by measuring their distance from nearby positions of the moon, whose longitude is then taken from computation.

For a fragment of a Babylonian star catalog cf. Sachs, J. of Cuneiform Studies 6 (1952) p. 146–150.

A model of a mechanical clock, based on stereographic projection, is described by Diels, Antike Technik, 3rd ed. (1924) p. 217 and Pl. 18. Stereographic projection is applied by Ptolemy in his "Planisphaerium" (opera II p. 225–259 Heiberg) in order to determine the rising times by using only plane trigonometry. The same construction is used for the "astrolabe", an instrument which allows us to represent the celestial phenomena by means of the rotation of a network of circles over a disc representing the plane of the equator upon which the celestial sphere is projected from the south pole. Cf. my paper "The Early History of the Astrolabe" in Isis 40 (1949) p. 240–256.

ad 67. The Babylonian value 27;33,20 days for the anomalistic month is also mentioned by Geminus (about 70 B.C.) in his Introduction to Astronomy 18. The derivation, however, given by Geminus is wrong. He says that 717 anomalistic months contain 19756 days, but from this it follows that one month equals 27;33,13,18, ... days and not 27;33,20. Geminus is here telescoping two different methods into one.

References for the publications quoted in the text (p. 165):
Schnabel, Zeitschr. f. Assyriologie 37 (1927) p. 35 and p. 60.
Knudtzon-Neugebauer, Bull. soc. roy. des lettres de Lund, 1946–1947,
p. 77–88.
Neugebauer, Danske Videnskab. Selskab, hist.-filol. Meddelelser 32, No. 2
(1949).
Thibaut-Dvivedi, The Pañchasiddhântikâ, the Astronomical Work of
Varâha Mihira. Benares 1889.
John Warren, Kala Sankalita, a Collection of Memoirs on the Various
Modes According to which the Nations of the Southern Parts of India Divide
Time. Madras 1825.
Warren had a predecessor in the French astronomer LeGentil who was
sent to India to observe the Venus transits of 1761 and 1769. He missed the first
because of the French-English war, the second because of clouds. He learned,
however, a great deal about Tamil astronomy and gave an excellent description
of the methods for computing eclipses in the Mémoires of 1772, II of the French
Academy. The French scholars of this period, Cassini, LeGentil, Bailly, Delambre,
had reached a clear distinction between the linear methods of the Tamil astronomers and the trigonometric type of the Sūrya Siddhānta. This insight has been
lost in the subsequent literature.
Translation of the Sûrya-Siddhânta by Ebenezer Burgess, reprinted 1935,
University of Calcutta, from J. Am. Oriental Soc. 6 (1860) p. 141–498.
For a discussion of the Tamil methods for the computation of lunar eclipses
cf. O. Neugebauer, Tamil Astronomy, a Study in the History of Astronomy
in India, Osiris 10 (1952) p. 252–276 and B. L. van der Waerden, Die Be-
wegung der Sonne nach griechischen und indischen Tafeln, S. B. Bayer. Akad.
d. Wiss., Math.-nat. Kl. 1952 p. 219–232 and by the same author Tamil Astronomy, Centaurus 4 (1956) p. 221–234.
There are many evident indications of a direct contact of Hindu astronomy
with Hellenistic tradition, e. g., the use of epicycles or the use of tables of chords
which were transformed by the Hindus into tables of sines. The same mixture of
ecliptic arcs and declination circles is found with Hipparchus (cf. p. 185) and
in the early Siddhāntas[1] (called "polar longitude" and "polar latitude" by
Burgess). The extensive use of the sexagesimal system is common to both
Greek and Mesopotamian astronomy. The use of "tithis", which are so characteristic of Hindu astronomy, is not yet attested in Greek texts but we know so
little about the linear methods in Hellenistic astronomy that we may assume
that the use of "lunar days" penetrated into Hellenistic astrology from Babylonian
texts exactly in the same fashion as the planetary periods and the lunar theory.[2]

---

[1]) Hipparchus divides not only the ecliptic but also the equator into 30° sections
and denotes them by the names of the zodiacal signs (cf. Manitius's edition of
Hipparchus's Commentary to Aratus p. 295). In the Sūrya Siddhānta, the zodiacal
signs are used in similar fashion to denote arcs on any great circle.
[2]) The use of "tithis", that is, of thirthieths of mean synodic months, was first
discovered in Babylonian planetary texts by Pannekoek (Koninklijke Akad.
van Wetensch. te Amsterdam, Proceedings 19, 1916, p. 684–703) and by van der
Waerden (Eudemus 1, 1941, p. 23–48). For the occurence of these units in Babylonian lunar ephemerides cf. O. Neugebauer, ACT I p. 119. In late Hindu astronomy

The occurrence of the ratio 3:2 for the longest and shortest days might be taken as a sign of direct Mesopotamian influence though also this element is a part of the Hellenistic tradition of the "climates". Also the arrangement of the planets according to the "rulers" of the days of the week (cf. p. 169) indicates primarily Hellenistic influence.

For the Roman sea-routes and for Roman settlements in India, see R. E. M. Wheeler, Arikamedu: An Indo-Roman Trading-Station on the East Coast of India, Ancient India, No. 2 (1946) p. 17–124. For a summary see Martin P. Charlesworth, Roman trade with India, Studies in Roman Economic and Social History in Honor of Allan Chester Johnson, Princeton 1951, p. 131–143. A translation, with extensive commentary, of the Periplus was published by W. H. Schoff, The Periplus of the Erythrean Sea, New York–Philadelphia 1912.

A relatively early date for Greek-Persian-Hindu contacts seems to be obtainable from a passage in the Dēnkart, Book IV, according to which Hindu works on grammar and on astronomy and horoscopy as well as the Greek Almagest reached the court of Shapur I (about 250 A.D.); cf. Menasce, Journal Asiatique 237 (1949) p. 2 f.

ad 68. Hipparchus is often quoted in the astrological literature. As an example might be mentioned Vettius Valens I, 19 for the elongation of the moon. This method uses the same epoch (Augustus — 1) as P. Ryl. 27 and is therefore not genuinely Hipparchian. Nevertheless it may go back to Hipparchus just as other linear methods were developed from Babylonian originals. It was F. Boll who first emphasized that the ancient reports connecting Hipparchus with astrology have to be taken seriously in view of the time of origin of astrological doctrine in the second century B.C. (cf. Boll's lecture "Die Erforschung der antiken Astrologie" in Neue Jahrbücher für das Klassische Altertum 21, 1908, p. 103–126). F. Cumont speaks about "Hipparque, dont le nom doit être placé en tête des astrologues comme des astronomes grecs" (Klio 9, 1909, p. 268).

ad 69. The earliest known horoscope is cast for the year 410 B.C. (A. Sachs, Babylonian horoscopes, J. of Cuneiform Studies 6 (1952) p. 49–75). The remaining cuneiform horoscopes belong to the Seleucid period. The earliest Greek horoscope is the horoscope of the coronation by Pompey of Antiochus I of Commagene in 62 B.C. on the Nimrud Dagh. Horoscopes on papyrus or in Greek literature start at the beginning of our era.

An early indication of knowledge of Babylonian astrology in Greece was pointed out to me by Professor H. Cherniss. Proclus (who died in A.D. 485) in his commentary to Plato's Timaeus (III, 151 Diehl) quotes Theophrastus, the successor of Aristotle (died 322 B.C.), as saying that the Chaldeans were able to predict, in his time, not only the weather from the heavens but also life and death of all persons.

the tithis have become of variable length, being thirtieths of the true lunar months. Thus one finally introduced the variability of these units which had been invented in order to avoid the fluctuations of the true lunar calendar. In the classical Hindu astronomy, however, (e. g., in the Sūrya Siddhānta) the tithis are of fixed mean length; cf. Olaf Schmidt, On the Computation of the Ahargana, Centaurus 2 (1952) p. 140–180.

Still one generation earlier leads to an oft-quoted remark of Cicero (De divinatione II, 42, 87) that Eudoxus has written that one should not believe the Chaldean practice of predicting the fate of a person from the date of his birth.

Though the possibility of an early spread of Babylonian astrology to the West cannot be denied, caution seems to me to be necessary. It is by no means certain that prediction "from the day of birth" means astronomical prediction. On the contrary, we have a very similar earlier reference by Herodotus, who says, (II, 82) that the Egyptians "assign each month and each day" to a god and that "they can tell what fortune, what end, and what disposition a man shall have according to the day of his birth". Here we see clearly that prediction from "the day of birth" means not at all prediction from the planetary positions and from the position of the zodiac at the hour of birth. As Brugsch has seen (Herodotus ed. Stein, 1881, p. 88 note) Herodotus is referring to a practice which is directly attested from texts, e. g., the P. Sallier IV of the British Museum, written in the New Kingdom (cf. F. Chabas, Le calendrier des jours fastes et néfastes de l'année égyptienne [1870]; Oeuvres diverses, vol. IV p. 127–235; also F. W. Read, Proc. Soc. Bibl. Arch. 38 [1916] p. 19–26, 60–69, and Abd el-Mohsen Bakir, Annales du Service des Antiquités de l'Égypte 48 [1948] p. 429). These lists of lucky and unlucky days contain also predictions as to the future fate of a person born on a certain day, e. g., death by a crocodile, snake bite, etc., exactly as Herodotus indicates.

A still more rudimentary form of prediction of the future of a person can be found in Hittite texts of the 13th century B.C. (cf. B. Meissner, Klio 19 (1925), p. 432–434), where the fate of a child is made dependent upon the month of its birth. This is, of course, fundamentally different from the planetary horoscopy of the Hellenistic age.

An important argument for the comparatively late introduction of astrology seems to me the frequent use of Aries 8° as vernal point in astrological texts, i. e. the vernal point of System B of the Babylonian lunar theory, which seems to have reached the Greeks only at the time of Hipparchus. Eudoxus, however, uses Aries 15° as vernal point and the same holds for several astronomical and calendaric papyri of the Ptolemaic period in Egypt. This earlier, Eudoxian norm appears nowhere in astrological literature.

There exists, moreover, direct evidence about the type of material that was associated with the name of Eudoxus. Bezold and Boll, Reflexe astrologischer Keilschriften bei griechischen Schriftstellern (S.B. Heidelberger Akad. d. Wiss. Philos.-hist. Kl. 1911 No. 7) have shown close parallels between Mesopotamian omens concerning thunder, clouds, the positions of the horns of the moon, etc., and Greek calendars of the same type which are also related to Eudoxus. But nowhere in these parallels does there occur any reference to computational methods which are characteristic for horoscopic astrology.

The techniques and doctrines of Greek astrology are described in the classical work of A. Bouché-Leclercq, L'astrologie grecque, Paris 1899. The historical and philosophical aspects of one of the major components of astrological lore and its branching out into botany, alchemy, etc. is discussed by A. J. Festugière,

La Révélation d'Hermès Trismégiste, I, L'astrologie et les sciences occultes, 2nd ed., Paris 1950.

For the Hellenistic origin of astrology see W. Kroll, Klio 18 (1923) p. 213 and W. Capelle, Hermes 60 (1925) p. 373.

For the history of the planetary week see F. H. Colson, The Week, Cambridge University Press 1926. Its penetration into Jewish literature has been investigated by S. Gandz, Proc. Am. Acad. for Jewish Research 18 (1949) p. 213–254.

ad 70. For Teukros cf. Gundel in Pauly-Wissowa, Real Enc. 5 A col. 1132–1134. An upper limit for his lifetime would be given by Antiochus of Athens, to whom Teukros is known (cf. Real Enc. 18, 3 col. 122, 58 ff.). Cumont (Annuaire de l'institut de philol. et d'hist. orientales 2, Bruxelles 1934, p. 135–156) places Antiochus between 100 B.C. and 50 A.D., but his arguments are of a very indirect character. Thus we have at best these same limits for Teukros.

Teukros is called "the Babylonian" by Porphyrius (about 270 A.D.) and by subsequent authors[1]. Gundel, who realized how little is actually known about Babylonian astrology, went to the other extreme by practically completely denying Babylonian influence on Hellenistic astrology and substituting Egyptian mythology as the main agent. Consequently he adopted a rather artificial hypothesis of Eisler that "Babylon" refers to the fortress-town of this name in Egypt (near Cairo). I see no evidence which supports this viewpoint.

The writings of Teukros are known only through excerpts preserved in later astrological treatises. Their great historical importance was first fully recognized by Boll in his "Sphaera, neue griechische Texte und Untersuchungen zur Geschichte der Sternbilder" (Leipzig, Teubner, 1903). For the translation into Pehlevi cf. Nallino, Raccolta di Scritti VI p. 285–303. The Arabic tradition was investigated by Steinschneider (Z. Deutsche Morgenländische Gesellschaft 50, 1896, p. 352–354). Cf. also Gundel, Real Enc. 5 A col. 1133.

For the dialogue "Hermippus" cf. Fr. Boll in S.B. Heidelberg Akad. d. Wiss., phil. hist. Kl. 1912, Abh. 18 and V. Stegemann, Hermes 78 (1944) p. 120 note 2.

Babylonian parameters and methods found in the Pañca Siddhāntika are quoted in my paper "Babylonian Planetary Theory" Proc. of the Am. Philos. Soc. 98 (1954), p. 60–89.

ad 71. Datta and Singh in their "History of Hindu Mathematics", vol. I (1935), quote (p. 59) the commentary of Bhaṭṭotpala to the Brihat Samhita of Varāha Mihira for an excerpt from the original Paulisa Siddhānta in which a huge number, ending in ..7800, is expressed by number words in opposite arrangement in the form "zero, zero, eight, seven, ...". It seems to me rather plausible to explain the decimal place value notation as a modification of the sexagesimal place value notation with which the Hindus became familiar through Hellenistic astronomy.

For the date of the Sūrya-Siddhānta cf. van der Waerden, Diophantische

---

[1] In case this name is based on ancient tradition one could not be sure that Teukros was a citizen of Babylon, because "Babylonian" is also used for inhabitants of Seleucia on the Tigris; cf. Tarn, The Greeks in Bactria and India, p. 15.

Gleichungen und Planetenperioden in der indischen Astronomie, Vierteljahrschrift der Naturforsch. Ges. in Zürich, 100 (1955) p. 153–170.

For Ibn Yūnus's reference to Persian observations cf. Caussin, Notices et
Extraits de Manuscrits de la Bibliothèque Nationale, tome 7, 1 and 12 [1803]
p. 234 note (1). For the date of Vettius Valens cf. O. Neugebauer, Harvard
Theological Review 47 (1954) p. 65–67.

An excellent summary of Greek magic can be found in K. Preisendanz,
Zur Überlieferungsgeschichte der spätantiken Magie, Zentralblatt für Bibliotheks-
wesen, Beiheft 75 (1951), p. 223–240.

In our discussions we have frequently used the word "Greek" with no further
qualification. It may be useful to remark that we use this term only as a convenient
geographical or linguistic notation. A concept like "Greek mathematics", however,
seems to me more misleading than helpful. We are fairly well acquainted with
three mathematicians—Euclid, Archimedes, and Apollonius—who represent one
consistent tradition. We know only one astronomer, Ptolemy. And we are
familiar with about equally many minor figures who more or less follow their
great masters. Thus what is usually called "Greek" mathematics consists of the
fragments of writings of about 10 or 20 persons scattered over a period of 600
years. It seems to me a dangerous generalization to abstract from this material
a common type and then to establish some mysterious deeper principle which
supposedly connects a mathematical document with some other work of art.

APPENDIX I

The Ptolemaic System.

74. The following is a description of the cinematic models according to which the tables of the Almagest were computed, allowing us to find for any given moment the longitudes of the sun, the moon and the five planets. This will give the reader at least some concept of the basic methods developed in the Almagest, though I cannot emphasize too strongly the fact that I am presenting here only one special facet of the planetary theory of the Almagest.

Even within the restricted problem of determining the coordinates of the celestial bodies, I have omitted the problem of latitudes because an adequate discussion could not be given without a detailed description of the empirical data which Ptolemy had assembled and of the geometrical and numerical procedures devised for their representation. Similarly I have almost completely ignored the very elegant methods which led from a great variety of empirical data to the numerical determination of the characteristic parameters of the cinematic models. Any serious student of ancient or medieval astronomy must familiarize himself with these details, not only in order properly to appreciate one of the greatest masterpieces of scientific analysis ever written, but also to be able to understand what was common knowledge for every competent astronomer from the second to the seventeenth century.

The planetary theory of the Almagest contains much more than the determination of longitude and latitude. An interesting section concerns the stationary points, another deals with first and last visibilities. The lunar theory is also concerned with parallax, distance and size of sun and moon, and the computation of eclipses. Observational instruments are described, mathematical tools devised. Solar apogee, equation of time, precession, fixed star coordinates and their connection with lunar positions are important questions which later became a center of interest for Islamic astronomers. These processes can be adequately described only in a work of a scale totally different from the present volume. I hope however, that the following description will be useful as a guide to the reading of the original sources.

75. The solar motion is represented in the Almagest by a simple eccentric motion, or by the equivalent epicyclic motion which depends only on one variable, the mean distance α from the apogee (cf. Fig. 31). The apogee is assumed to be fixed at a distance of 65;30° from the vernal point[1]), the eccentricity is $e = r = 2;30$ (for a deferent radius 60). The corresponding correction δ, called "prosthaphairesis for anomaly" is negative for values of the anomaly $\gamma = \alpha$ between 0 and 180 and positive for the remaining semicircle; its maximum value is 2;23° by which the "mean sun" at C can differ from the "true sun" at S. The corrections δ as function of the anomaly are given in a table in Almagest III,6 which is represented graphically in Fig. 33 (lowest curve[2])). This correction was later called the "equation of center".[3])

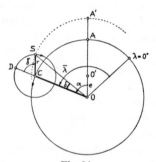

Fig. 31.

76. The lunar theory obviously requires a model of greater complication since it was known already to the Babylonian astronomers of the third or fourth century B.C. that the anomalistic month (return to the same velocity) was longer than the sidereal

---

[1]) In other words, it is assumed that tropical and anomalistic year coincide.
[2]) Representing the absolute value of δ.
[3]) To my knowledge, the term "*equatio*" appears first in Latin translations of Arabic treatises.

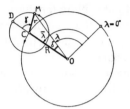

Fig. 32.

month (return to the same fixed star). Consequently in an epicyclic model (cf. Fig. 32) the mean motion in longitude, expressed by the angle $\bar{\lambda}$, is greater than the mean motion in anomaly, the first being about $13;10,35°^{/d}$ the latter $13;3,54°^{/d}$. The resulting motion can also be described as an eccentric motion with a rotating apsidal line. Measured in the units for which the radius of the deferent is 60, the radius of the epicycle was found to be $r = 5;15$. The resulting corrections $\delta = \lambda - \bar{\lambda}$ which leads from the "mean longitude" $\bar{\lambda}$ to the "true longitude" λ is tabulated in Almagest IV,10 and graphically represented as c_4 in Fig. 33 in the same scale as the equation of center of the sun.

77. The theory described so far was known to Hipparchus though refined in several respects by Ptolemy. The determination of the essential parameters was based on carefully selected observations of lunar eclipses and the results obtained were very satisfactory for the description of eclipses in general. Ptolemy, however, realized from a masterful analysis of observations of his own and of his predecessors that marked deviations from the predicted longitudes of the moon, reaching a maximal amount in the neighborhood of elongations of $\pm 90°$ from the sun, could occur. In other words he realized that the traditional lunar theory agreed with observations in the syzygies (conjunctions and

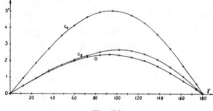

Fig. 33. *

oppositions) but could not explain longitudes near the quadratures, particularly for values of the anomaly γ near $\pm 90°$. In these cases the diameter of the epicycle seemed to be enlarged over the value found at the syzygies.

The procedure which Ptolemy followed to cope with this situation is of interest in many respects. It provides us with a

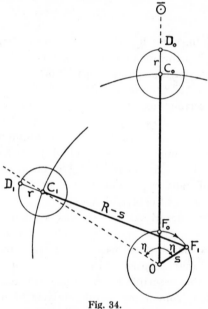

Fig. 34.

good insight into the mathematical and astronomical methodology of the time; the attitude toward a glaring defect of the theory is very revealing and has repercussions in Islamic astronomy and in the work of Copernicus; finally the method by which this inequality of the lunar motion was accounted for influenced also the planetary theory, both of Ptolemy and Copernicus.

As we have said above the observations suggested a dependence of the apparent diameter of the epicycle of the moon on the elongation from the sun. Such an effect could be produced by bringing the epicycle closer to the observer by the following mechanism (Fig. 34). Let C_0 be the position of the center of the lunar epicycle at conjunction with the mean sun such that OF_0C_0

are on a straight line and $OC_0 = R$ the radius of the deferent known from the previous model given in Fig. 32. Let η be an angle increasing proportionally with time at a rate equal to the difference between the mean velocity of the moon and the mean velocity of the sun; η is therefore called the "elongation". Its value is zero at conjunction; as η increases, the point F is made to move in retrograde direction on a small circle of radius s and with center O such that its angular distance from the direction from O to the mean sun equals the elongation η. At the same time the point C, the center of the epicycle, moves forward such that the direction OC makes an angle η with the direction from O to the mean sun. In this way C approaches O as η increases from $0°$ toward $90°$. At quadrature ($\eta = 90$) the distance C from O reaches its minimum value $R - 2s$. At opposition ($\eta = 180$) C is again removed to the original distance R from O. Since eclipses can only occur at conjunctions or oppositions, the new model agrees with the old one for all the elements obtained from eclipses. Toward the quadratures, however, it increases the apparent diameter of the epicycle in accordance with the observations.

Ptolemy found one more deviation from the original theory for positions of the epicycle at elongations nearer to the octants: instead of counting the anomaly γ which determines the distance of the moon M on the epicycle from the apogee D, it had to be measured from a variable apogee H (Fig. 35) such that the radius HC has a "direction" (Greek: πρόσνευσις) toward a point N which is always located diametrically opposite to the point F. The point H is called the "mean apogee" because from it is measured the mean anomaly γ. The point T, which is the apogee of the epicycle as seen from O, is then called the "true" apogee.

The parameters of this model are

$$r = 5;15 \qquad s = 10;19 \qquad R = 60^1)$$

which show that the moon at quadrature could come as near as $R - 2s - r = 34;7$ to the observer. This obviously means that the apparent diameter of the moon itself should reach almost twice its mean value which is very definitely not the case. This discrepancy is silently ignored by Ptolemy, though he could not

[1]) In the "Canobic Inscription" (Opera II p. 150, 5 and 12), Ptolemy adopts the norm $R - s = 60$ and consequently obtains $r = 6;20$ $s = 12;28$.

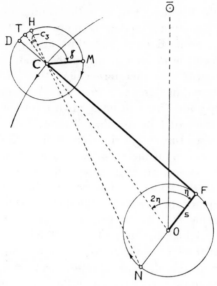

Fig. 35.

have doubted that the actual geocentric distances of the moon
were very different from what his model required. Nevertheless
this model was retained by almost all his followers simply because
it proved to predict at least the longitudes correctly.

Copernicus pointed to the obvious discrepancy between
Ptolemy's lunar model and the observable parallaxes[1]) and
proposed another model which would keep the center C of the
epicycle at mean distance but would nevertheless increase the
moon's distance from C at quadratures (Fig. 36). He assumed
that the moon M was located on a secondary epicycle such that
it started its motion at conjunction at E at a distance $r - s$ from C.
With increasing elongation η, the moon would move on the small
epicycle in the direct sense by 2η, thus reaching for $\eta = 90$ a
distance $r + s$ from C. Since Copernicus used $r = 6;35$ and
$s = 1;25$ for $R = 60$ the moon could not come closer to the
observer than $R - (r + s) = 52$ which is no longer essentially
different from the distances resulting from the simple model
(Fig. 32).

[1]) De revolutionibus (published 1543) IV, 2.

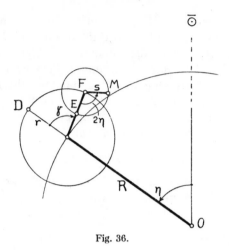

Fig. 36.

This obvious advantage of the use of secondary epicycles induced Copernicus to apply the same construction also to the planetary motion and thus to initiate complications which destroyed the inherent elegance and simplicity of the Ptolemaic model.

Only recently has it been discovered[1]) that the same method for the correction of Ptolemy's lunar model was used about 200 years before Copernicus by ibn ash-Shātir. Whether Copernicus knew about his predecessor or not is impossible to decide at the present moment.

78. The procedure for computing the longitude of the moon according to the Ptolemaic model is based on the tables in Almagest V,8. The first two columns contain the argument Θ and $360 - \Theta$ for Θ between $0°$ and $180°$. The 3rd column gives the angular difference between true and mean apogee of the epicycle (T and H in Fig. 35) as function of the double elongation 2η, where η is the difference between mean lunar and mean solar longitude at the given moment. We call this function $c_3(2\eta)$; its graph is given in Fig. 37. By forming $\gamma' = \gamma + c_3$ one has found the anomaly which determines the angle at which M is seen from O. With γ' as argument one finds in the 4th column the value $c_4(\gamma')$ which is the equation of center already known from

[1]) Cf. V. Roberts, in a forthcoming article in Isis.

the simple lunar model (Fig. 32 and 33), that is, the angle under which CM would be seen from O if the distance of C were $R = 60$. The column $c_5(\gamma')$ gives the amount by which the equation of center increase if C were at minimum distance from O; finally $c_6(2\eta)$ indicates the fraction of c_5 due to the fact that for $2\eta < 180°$ the actual distance of C from O lies between the two extrema for which the equation of center is c_4 and $c_4 + c_5$ respectively. Thus one finds for the final equation the value

$$\delta = c_4 + c_5 \cdot c_6$$

and hence for the true longitude $\lambda = \overline{\lambda} - \delta$ if $\gamma' < 180$ or $\lambda = \overline{\lambda} + \delta$ if $\gamma' > 180$. The numerical values given in Almagest V,8 are graphically represented in Figs. 33 and 37.

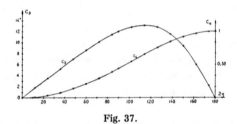

Fig. 37.

79. Ptolemy's planetary theory follows the same principles as the solar and lunar theory. As we have seen (p. 123 f.), circular planetary orbits with the sun as center are epicyclic motions with respect to the terrestrial observer. It is easy to see by the same method that eccentric circular orbits lead to epicyclic motions with eccentric deferents, the eccentricity being the resultant of the vectors which represent the planetary and solar (or terrestrial) eccentricity.

This model was further modified by Ptolemy on the basis of observations mostly made by himself or his predecessor Theon;[1]) he found that the center of the epicycle appears to move with its mean angular velocity not with respect to the center of the deferent but with respect to a point (later called "equant") located symmetrically to the observer. We have encountered essentially the same idea in Ptolemy's theory of the moon where the center C

[1]) Perhaps Ptolemy's teacher, not to be confused with Theon of Alexandria, the author of the "Handy Tables" who lived two centuries after Ptolemy.

of the epicycle is moving with constant angular velocity only in so far as it is kept on the radius OT which moves uniformly (Fig. 35) while the linear velocity of C is not constant. Philosophical minds considered this departure from strictly uniform circular motion the most serious objection against the Ptolemaic system and invented extremely complicated combinations of circular motions in order to rescue the axiom of the primeval simplicity of a spherical universe.

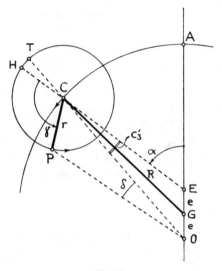

Fig. 38.

80. Fig. 38 represents Ptolemy's model for the outer planets. The mean motion of the planet is represented by the motion of the center C of the epicycle, measured by the "mean distance" α of C from the apogee A. The planet P moves on the epicycle with a speed corresponding to the synodic period and measured by the "anomaly" γ. The sense of rotation of P on the epicycle is now equal to the sense of mean motion, thus giving the planet its greatest direct motion near the apogee of the epicycle and producing retrogradation near the perigee.

In agreement with our general analysis of heliocentric and geocentric motion (p. 124 Fig. 14) the direction CP is parallel to the direction from O to the mean sun in the case of an outer planet. In the case of an inner planet the angle α increases as the longitude of the mean sun (in fact C may be identified with the mean sun) and the anomaly γ varies independent of the position of the sun. In all cases the position of P depends on the two independent variables α and γ which can be considered to be known for any given moment t.

For Mercury the observations could not be reconciled with so simple a model as for the other planets—"Nulle planète n'a demandé aux astronomes plus de soins et de peines que Mercure, et ne leur a donné en récompense tant d'inquiétudes, tant de contrariétés" says Leverrier. Ptolemy's data led again to the necessity of increasing the apparent size of the epicycle as in the case of the moon, the only difference being that the closest approach now occurred at about $\pm 120°$ from the apsidal line. Thus Ptolemy adopted a model as described in Fig. 39. The center G of the deferent moves retrograde on a circle of radius e and center B where e is not only the eccentricity BE but also the distance of the observer O from the equant E. The center C of the epicycle moves forward such that its progress, seen from E, appears uniform and of the same amount as G is removed in the

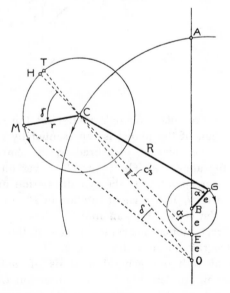

Fig. 39.

opposite direction from the apsidal line OA. For $\alpha = \pm 120$ the radius GC of the deferent passes through E and thus brings C nearer to O than for $\alpha = 180$. In other words the orbit of C with respect to O has one apogee in the apsidal line but two perigees symmetric to it at $\alpha = \pm 120$.

81. The practical computation of planetary positions follows much the same lines as for the moon. Almagest XI,11 gives tables in 8 columns for each planet. Columns 1 and 2 contain the common arguments Θ and $360 - \Theta$. Columns 3 and 4 are only used in the combination,

$$c'_3(\alpha) = c_3(\alpha) + c_4(\alpha)$$

for identical values of the mean distance α in both columns. In later works—Theon's tables and Islamic tables—these two columns are always combined into one (c'_3 in Fig. 38), leading from the mean apogee H to the true apogee T. Ptolemy kept the two columns separate for didactic reasons because he wanted the reader to see how c'_3 had been obtained from first locating the center of the deferent at E and then moving it to G.

With the argument $\gamma' = \gamma + c'_3$ three tables are computed: $c_6(\gamma')$ gives the angle $\delta(\gamma')$ under which the radius $r = CP$ of the epicycle appears from O under the assumption that C is located at mean distance from O. The column $c_5(\gamma')$ gives the amount of the correction which must be subtracted from c_6 such that $c_6 - c_5$ represents $\delta(\gamma')$ for the case of C at maximum distance from O. Similarly $c_6 + c_7$ gives $\delta(\gamma')$ for minimum distance of C from O. Finally $c_8(\alpha)$ gives the coefficient by which c_5 or c_7, respectively, have to be multiplied in order to give the correction of c_5 at a distance α of C from the apogee.

Now we can formulate the whole procedure for the computation of the true longitude λ of a planet. For the given moment t the following elements are known:

λ_0 longitude of the apogee A

α mean distance of C from A

γ mean anomaly of the planet.

Then the true longitude $\lambda = \lambda(t)$ is given by

$$\lambda = \lambda_0 + \alpha - c'_3(\alpha) + \delta$$

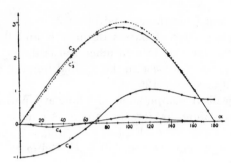

Fig. 40 a.

where $c'_3(\alpha)$ is found from

$$c'_3(\alpha) = c_3(\alpha) + c_4(\alpha)$$

and δ from

$$\delta = c_6(\gamma') + c_8(\alpha) \cdot \begin{cases} c_5(\gamma') & \text{for} \quad c_8 \leqq 0 \\ c_7(\gamma') & \text{for} \quad c_8 \geqq 0 \end{cases}$$

with

$$\gamma' = \gamma + c'_3(\alpha).$$

Fig. 40 represents these functions in the case of Mercury. The corresponding curves for the other planets are somewhat simpler since they have only one perigee of the center of the epicycle. The technique of computation, however, is in all cases the same.

82. It is illuminating to compare Ptolemy's model of the motion of Mercury with the Copernican theory. The empirical data are, of course, essentially the same, particularly the fact that the smallest values for the greatest elongation of Mercury from the

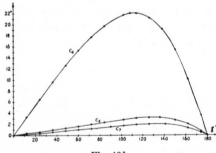

Fig. 40 b.

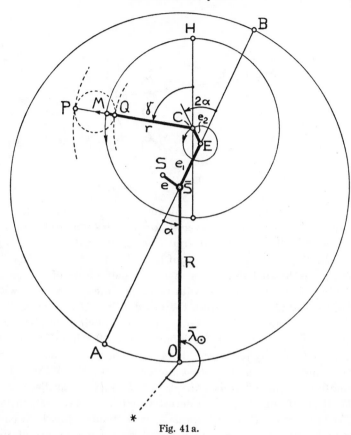

Fig. 41 a.

sun occur in Libra whereas the greatest values are not observed opposite to Libra in Aries but two signs before or after, in Aquarius and Gemini. In order to account for this observation, Mercury is made to move on a straight line segment such that its distance from the center of its orbit varies with the proper period. A movement on a straight line seems not quite in conformity with the postulate of circular motions of the celestial bodies but fortunately Copernicus had at his disposal a device of aṭ-Ṭūṣī, who had shown that a point of the circumference of a circle of radius $\frac{s}{2}$ moves along the diameter of a circle of radius s inside of which the first circle rolls (cf. Fig. 41 b).

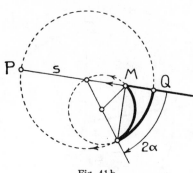

Fig. 41 b.

Fig. 41 a describes the Copernican modification of Ptolemy's theory of Mercury. The observer O now moves on a circle around the mean sun $\bar{S}$, the physical sun S being at a distance e from $\bar{S}$ corresponding to the eccentricity of the solar orbit[1]). At a distance e_1 from $\bar{S}$ is located the center E of a small circle of radius e_2 on which rotates the point C with twice the angular velocity with which the distance α of O from the apsidal line $\bar{S}A$ increases. The point C is the center of the instantaneous orbital circle of Mercury. Its radius $r = CM$ is made variable between the limits P and Q by means of aṭ-Ṭūsī's device (Fig. 41 b) such that M moves from Q to P and back again while α increases from $0°$ to $180°$. The anomaly γ is counted from a direction CH which is parallel to $\bar{S}O$.

83. It is, of course, of no interest whether we say that, in the model of Fig. 41 a, O rotates about S or $\bar{S}$ about O, in which case R is called the radius of the deferent and r the radius of the epicycle. Thus it is evident that cinematically the two models are hardly different except for Copernicus's insistence on using circles for every partial motion where Ptolemy had already reached much greater freedom of approach.

The popular belief that Copernicus's heliocentric system constitutes a significant simplification of the Ptolemaic system is obviously wrong. The choice of the reference system has no effect whatever on the structure of the model, and the Copernican models themselves require about twice as many circles as the Ptolemaic models and are far less elegant and adaptable. In fact

---

[1]) Copernicus is not very outspoken about the question whether $\bar{S}$ rotates about S or vice versa. For his planetary theory only $\bar{S}$ is of real significance.

the *Almagest*, al-Battānī's *Opus astronomicum* and Copernicus's *De revolutionibus*. Chapter by chapter, theorem by theorem, table by table, these works run parallel. With Tycho Brahe and Kepler the spell of tradition was broken. The very style in which these men write is totally different from the classical prototype. Never has a more significant title been given to an astronomical work than to Kepler's book on Mars: "*Astronomia Nova*".

BIBLIOGRAPHY TO APPENDIX I

For the history of Greek and medieval astronomy three works must be consulted: Delambre, Histoire de l'astronomie ancienne (2 vols., Paris 1817) and his Histoire de l'astronomie du moyen age (Paris 1819). Norbert Herz, Geschichte der Bahnbestimmung von Planeten und Kometen (2 parts, Leipzig, Teubner, 1887 and 1894). J. L. E. Dreyer, History of the Planetary Systems from Thales to Kepler (Cambridge 1906, reprinted New York 1953 under the title "A History of Astronomy").

See furthermore Norbert Herz in the Handwörterbuch der Astronomie vol. I p. 1–98 (= Encyklopaedie d. Naturwiss. III, 2 ed. W. Valentiner, Breslau 1897) and the article on Ptolemaios in Pauly-Wissowa, Realencyclopädie der classischen Altertumswissenschaft vol. 23, 2 cols. 1788–1859 by B. L. van der Waerden; also Paul Tannery, Recherches sur l'histoire de l'astronomie ancienne, Paris 1893. A good summary of the Ptolemaic planetary theory is given in Derek J. Price, The Equatorie of the Planetis, Cambridge Univ. Press 1955 p. 93–117.

NOTES AND REFERENCES TO APPENDIX I

ad 77, simultaneously a contribution to the medievalism in modern scholarship. Pierre Duhem in vol. I of his Système du Monde (1914) p. 494 f. has given a description of Ptolemy's lunar theory according to which the moon would become retrograde each month since he gave it the wrong sense of rotation (l. c. Fig. 11). It is also Duhem's opinion that the retrograde motion of the diameter FN (using my figure 35) represents "évidemment" the retrograde motion of the nodal line (thus producing every month a total lunar and solar eclipse!). This flagrant nonsense has now been repeated for some 40 years by the excerptors of Duhem's work.

Duhem's total ignorance of Ptolemy's lunar theory is a good example of the rapid decline of the history of science. The details of the ancient theory of the moon has been particularly well known in France since it was the object of discussion which occupied the French academy from 1836 to 1871 and in which Sédillot, Biot, Arago, Damoiseau, Libri, Munk, Reinaud, de Slane,

the importance of Copernicus's work lies in a totally differe
direction than generally announced. One may enumerate h
main contributions as follows:

(a) The return to a strictly Ptolemaic methodology which made
all steps from the empirical data to the parameters of the model
perfectly clear and opened the way to a refinement of the basic
observations which eventually led to the proper generalization of
the Ptolemaic methods, discarding the iterated epicycles of
Copernicus.

(b) The insight that we can obtain information about the
actual planetary distances if we assume that all planetary orbits
have essentially the same center, namely, the sun. Then the radii
of the epicycles of the inner planets directly give their distances
from the sun in terms of R; for the outer planets the reciprocals
of the radii of the epicycles measure the heliocentric distances.
Again the question which body is supposedly at rest is of no
interest whatever and therefore Tycho Brahe's system was
exactly as good as the Copernican system, and equally complicated
because of the same doctrine of circularity of admissible motions.

(c) The assumption of a common center of the planetary orbits
suggested also the proper solution of the problem of latitudes,
namely, that the inclined planes of the planetary orbits pass
through that common center. Unfortunately the postulate of
circularity induced Copernicus to use the mean sun and not the
real sun as common center and thus resulted in a theory of
latitudes which labored under exactly the same complications as
Ptolemy's theory. Nevertheless, this modification of the ancient
theory of latitudes helped Kepler to find the real solution, which
then permitted the computation of heliocentric coordinates in a
uniform fashion and the finding of the geocentric coordinates
through an independent procedure.

The enormous increase of empirical data accumulated by
Tycho Brahe and his collaborators finally convinced Kepler that
even the return to the Ptolemaic model with equant could not
properly represent the observations and thus led him to abandon
the axiom of circular orbits and to the discovery of the proper
orbits.

There is no better way to convince oneself of the inner coherence
of ancient and mediaeval astronomy than to place side by side

Chasles, Leverrier, Bertrand and others participated.[1]) Sédillot thought he had discovered in a passage of Abū'l-Wēfa (who died 998) a description of the lunar inequality, now called "variation" and published by Tycho Brahe in 1602 (Opera II p. 100 f.) whereas Ptolemy's theory covered only the first and second inequality (called "evection"). It is no longer possible to doubt that Sédillot was in error and Abū'l-Wēfa gave only a description of Ptolemy's construction without any addition of his own. It is, however, of interest to remark that P. Tannery has shown (Astronomie ancienne p. 211 ff.) that the combined effects of Ptolemy's theory do not exactly correspond to the "evection" of the modern theory of the lunar motion but also include about half of the term which corresponds to the "variation".

ad 80. The parameters for the planetary orbits are for R = 60

for Saturn	$r =$	6;30	$e =$	3;25
Jupiter		11;30		2;45
Mars		39;30		6;0
Venus		43;10		1;15
Mercury		22;30		3;0

None of my figures for the planetary models is drawn to scale since most details would become unrecognizable in the available space.

In contrast to the solar theory, in which the apogee is of constant tropical longitude, the apogees of the planetary orbits are assumed fixed with respect to the fixed stars and thus participate in their increase in longitude by 1° per century.

ad 82. Nāṣir ad-Dīn aṭ-Ṭūsī (born 1201, died 1274) was the director of the famous observatory at Marāgha. He objected on philosophical grounds against the crank-mechanism in Ptolemy's theory of the moon and of Mercury. The relevant chapter from his "Memento on Astronomy" was translated by Carra de Vaux in P. Tannery's Recherches sur l'histoire de l'astronomie ancienne, Paris 1893, p. 348–359. I do not know through what medium Copernicus knew about Ṭūsī's construction.

The motion of M on the diameter PQ (Fig. 41 b) is, of course, nothing but a simple harmonic motion $QM = s(1 - \cos 2\alpha)$.

The parameters, as determined by Copernicus (de revol. V, 27) are

$R = 10.000$	or, if $R = 60$:
$e = 322$	$e = 1;56$
$e_1 = 736$	$e_1 = 4;25$
$e_2 = 212$	$e_2 = 1;16$
$3573 \leq r \leq 3953$	$21;26 \leq r \leq 23;43$
$2s = 380$	$2s = 2;16$

Longitudes are counted by Copernicus sidereally, taking γ Arietis as zero point. This corresponds to a widespread mediaeval custom counting longitudes from Regulus (α Leonis), a norm which goes back at least to Ptolemy's Canobic Inscription (opera II p. 152, 2 f. to be corrected by p. 80, 27).

---

[1]) Cf. the excellent summary by Carra de Vaux, J. Asiatique 8 sér., 19 (1892) p. 440–471.

APPENDIX II

On Greek Mathematics

84. I do not consider it as the goal of historical writing to condense the complexity of historical processes into some kind of "digest" or "synthesis". On the contrary, I see the main purpose of historical studies in the unfolding of the stupendous wealth of phenomena which are connected with any phase of human history and thus to counteract the natural tendency toward over-simplification and philosophical constructions which are the faithful companions of ignorance.

To a modern mathematician who wants to get some insight into the mechanism of Greek mathematics, the access is made easy. The major works of Archimedes, Apollonius, Euclid, etc., are well edited and competently translated. The careful reading of a treatise by Archimedes or a book of Apollonius takes time and effort—as does the pursuit of all worthwhile knowledge—but one is repaid with a much deeper understanding of ancient mathematical methods than the reading of all summaries could provide. After a solid basis has been established, works like T. L. Heath's will provide a competent guide to related material and give the general background.

The astronomical material is much less directly accessible and consequently certain mathematical methods which were developed in close relationship to astronomy are much less generally known than, e. g., Euclid's procedures. The conventional picture of Greek mathematics—a very sophisticated branch of geometry followed by some not quite successful attempts at algebra and number theory during the later periods of decline—is wrong for two reasons. First, as we have seen on many occasions in the preceding chapters, we are not dealing with a decline from scientific geometry to less exact methods of algebraic tendency, but with two contemporary phenomena: a comparatively rapid development of rigid mathematical reasoning on the highest level in contrast to a much older and little changing background of ancient oriental and Hellenistic mathematics with a strongly algebraic character inherited from its Babylonian origin. Secondly, even the strictly Greek development is not adequately described by the Euclidean-Archimedean development, which is most familiar to the modern reader, but we must add many methods which concern numerical and graphical problems which originated in mathematical astronomy. In fact we see here the first instance of the stimulus which astronomy has repeatedly given to mathematics. It is the purpose of the following to illustrate some of these less known aspects of Greek mathematics.

85. The fact that Greek plane trigonometry was based on the tabulation of chords instead of sines produces, of course, no essential differences between modern and ancient procedure. The norm $R = 60$ introduces coefficients $1/120$ but since all fractions are written sexagesimally this only implies division by 2,0 or numerically the same as halving. Thus we have (Fig. 42)

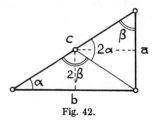

Fig. 42.

$$a = \frac{c}{2,0}\, crd\, 2\alpha \qquad\qquad a = c \sin \alpha$$

$$b = \frac{c}{2,0}\, crd\, (180 - 2\alpha) \qquad b = c \cos \alpha$$

$$\frac{a}{b} = \frac{crd\, 2\alpha}{crd\, (180 - 2\alpha)} \qquad \frac{a}{b} = \tan \alpha$$

Since the tables extend from 0 to 180 the first two cases are just as easy to handle as in our system. The only real inconvience lies in the lack of tables for the ratios corresponding to tan α.

As an example for the use of trigonometry in an astronomical problem, I shall describe the method followed in Almagest IV, 6 for finding the length of the radius r of the lunar epicycle and the position of the apogee. We are dealing here with the simple model of the lunar motion (as in Fig. 32 p. 193) in which the center C of the epicycle moves on a fixed deferent of radius $R = 60$. This problem is of interest for several reasons. It represents a method which was certainly used and probably invented by Hipparchus. It is the simplest case of a much more general problem, namely, to determine the parameters of an orbit from a set of observed positions. It has, finally, close relations to an important problem in geodesy, namely, to find the position of an observer with respect to three given points.

In the case of the moon the essential steps are as follows. Observation of the motion of the moon with respect to the stars from day to day easily reveals that its velocity is not constant. About once every month this progress is at a minimum of about 12° and then again at a maximum of about 14° per day. Consequently it is easy to count the number of periods of the lunar velocity corresponding to a given number of lunar months. Within a few years of observations the mean length of this "anomalistic" period can be determined with sufficient accuracy. If we then decide to describe this variation of velocity by means of an epicyclic model, as in Fig. 32 (p. 193), we can consider as known the rate of change of the "anomaly" γ as well as the rate of change of the mean longitude λ of the center of the epicycle. The problem now faced consists in the determination of r and of the moment at which the moon is exactly at its minimum speed, i. e., in the apogee of the epicycle.

To this end the moments and longitudes of 3 lunar positions are determined by means of 3 lunar eclipses the longitudes of which are accurately known through the diametrically opposite solar positions. The reason for this procedure is very characteristic: one reaches greater accuracy by means of a computed solar position than from a direct comparison of the position of the moon with respect to stars, since the coordinates of stars would involve the measurement of angles by means of instruments the accuracy of which is difficult to control.

The three moments for the midpoints of the eclipses and the three corresponding longitudes furnish us with two sets of differences Δt_1, Δt_2, and $\Delta \lambda_1$, $\Delta \lambda_2$ (cf. Fig. 43 for the first pair of posi-

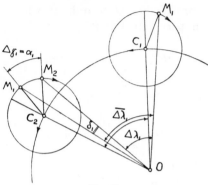

Fig. 43.

tions). Our knowledge of the mean values for the change of γ and $\bar{\lambda}$ furnishes us, in combination with $\varDelta t_1$ and $\varDelta t_2$, with the values $\varDelta \gamma_1$, $\varDelta \gamma_2$ and $\varDelta \bar{\lambda}_1$, $\varDelta \bar{\lambda}_2$ by which γ and $\bar{\lambda}$ have changed between consecutive eclipses. Fig. 43 illustrates the situation for the first two eclipses. From the data which we have so far assembled, we know that the observer at O sees the segment $M_1 M_2$ of the epicycle under the angle $\varDelta \bar{\lambda}_1 - \varDelta \lambda_1$ whereas this same segment is subtended at the center C of the epicycle by the angle $\varDelta \gamma_1$. A similar situation holds for the second pair of eclipses and this provides us with the final formulation of the problem (Fig. 44):

Three points M_1, M_2, M_3 on a circle of radius r subtend at its center given angles α_1, α_2 and are seen from an observer in O under given angles δ_1, δ_2. Find r and the position of O with respect to M_1, M_2, M_3.

This problem is solved as follows (Fig. 44): consider the

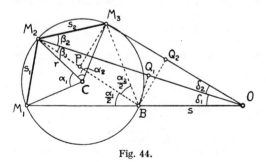

Fig. 44.

point B on the straight line OM_1 and call s the distance OB. We then have for the perpendicular BQ_1:

$$BQ_1 = \frac{s}{120} \, crd \, 2\,\delta_1$$

and also, because $\quad \beta_1 = \dfrac{\alpha_1}{2} - \delta_1$

$$BQ_1 = \frac{M_2 B}{120} \, crd \, 2\,\beta_1$$

and thus:

$$M_2 B = s \, \frac{crd \, 2\,\delta_1}{crd \, 2\,\beta_1}.$$

Similarly, since $\beta_1 + \beta_2 = \dfrac{\alpha_1 + \alpha_2}{2} - (\delta_1 + \delta_2)$

$$M_3 B = s \, \frac{crd \, 2(\delta_1 + \delta_2)}{crd \, 2(\beta_1 + \beta_2)}.$$

From the triangle $M_2 M_3 B$ we can now express the chord s_2 in terms of s and the given angles. Using the altitude $M_3 P$ we have

$$M_3 P = \frac{M_3 B}{120} \, crd \, \alpha_2 = \frac{s}{120} \cdot \frac{crd \, 2(\delta_1 + \delta_2)}{crd \, 2(\beta_1 + \beta_2)} \cdot crd \, \alpha_2$$

and

$$BP = \frac{M_3 B}{120} \, crd \, (180 - \alpha_2) = \frac{s}{120} \cdot \frac{crd \, 2(\delta_1 + \delta_2)}{crd \, 2(\beta_1 + \beta_2)} \cdot crd \, (180 - \alpha_2).$$

Consequently

$$M_2 P = M_2 B - BP = s \cdot (\,.\,.\,.\,.\,)$$

is known in terms of s and the given angles and therefore also

$$s_2 = \sqrt{M_2 P^2 + M_3 P^2} = s \cdot (\,.\,.\,.\,.\,.\,)$$

On the other hand s_2 is a chord of the circle of center C, thus

$$s_2 = \frac{r}{60} \, crd \, \alpha_2$$

where α_2 is given. Thus

$$r = \frac{60}{crd \, \alpha_2} \cdot \sqrt{M_2 P^2 + M_3 P^2} = s \cdot (\,.\,.\,.\,.\,.\,)$$

is known if s is known.

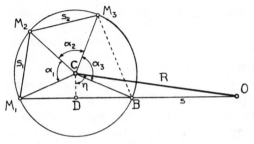

Fig. 45.

Obviously the absolute dimensions of Fig. 44 are not determined by the angles alone. Thus we must arbitrarily fix some distance and this is done by defining the radius OC of the deferent as $R = 60$. All other distances will then be measured in these units.

In order to find s we have to determine the remaining angles around C (cf. Fig. 45). Now

$$M_3 B = \frac{r}{60} \, crd \, \alpha_3$$

where both $M_3 P$ and r are expressed by trigonometric functions multiplied by s. Thus s can be cancelled in the above relation and α_3 is known independently of s. The same holds for

$$\alpha_4 = 360 - (\alpha_1 + \alpha_2 + \alpha_3).$$

Now s can be determined as follows

$$BM_1 = \frac{r}{60} \, crd \, \alpha_4 \quad \text{and} \quad OM_1 = s + BM_1 = s \cdot (\ldots).$$

Using a classical theorem which holds for a circle and a point outside, we have

$$(R + r) \cdot (R - r) = OM_1 \cdot s$$

or

$$R^2 = r^2 + OM_1 \cdot s = s^2 \cdot (\ldots).$$

Thus s is known in units of $R = 60$ and therefore also r.

Finally (Fig. 45)

$$OD = \frac{R}{120} \, crd \, 2\eta = \frac{1}{2} \, crd \, 2\eta = s + \frac{1}{2} BM_1$$

and thus η is known. Thus the moon M_1 at the moment of the first eclipse was on the epicycle $\dfrac{\alpha_4}{2} + \eta$ distant from the line OR. This completes the solution of our problem.

86. We have very little knowledge about the early history of spherical trigonometry. Beginning with quite primitive treatises on spherical astronomy by Euclid and Autolycus (4th century B.C.) we have several works which precede Menelaos (about 100 A.D.) in which one can recognize attempts to solve in general problems of astronomical importance, e. g., the determination of the rising times of given arcs of the ecliptic (cf. p. 160). We do not, however, know how numerical problems of this type were solved in practice; one might, e. g., assume that Hipparchus used methods which are known to us from Hindu astronomy and, in certain traces, even from the Almagest. These methods are characterized by the use of the interior of the sphere for the determination of the length of circular arcs on the sphere. For example the Sūrya Siddhānta (II, 60/63) determines the length of daylight as follows. Assume to be given the length s_0 of the equinoctial noon shadow of a gnomon of length 12 and the declination δ of the sun for the day in question. Then we have (Fig. 46)

$$\frac{12}{s_0} = \frac{OB}{e} = \frac{\mathrm{Sin}\ \delta}{e}$$

where we define $\mathrm{Sin}\ \delta = R \sin \delta$ according to the Hindu usage of trigonometric functions. Furthermore, r, the radius of the parallel circle of declination δ, the so-called "day radius" is given by

$$r = R - \mathrm{Sin\ vers}\ \delta$$

where the versed sine corresponds to our $1 - \cos \delta$. Finally

$$\mathrm{Sin}\ \alpha = R\frac{e}{r} = \frac{R\ \mathrm{Sin}\ \delta}{12(R - \mathrm{Sin\ vers}\ \delta)} \cdot s_0$$

where α is called the "right ascensional difference".

Using the value of α thus obtained the length of daylight is given by $180 + 2\alpha$ degrees.

87. The above approach to the solution of problems of spherical geometry by means of plane trigonometry applied to properly

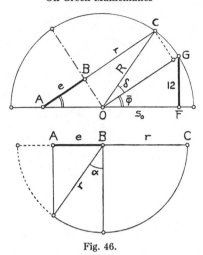

Fig. 46.

chosen planes is systematically expanded in the theory of the
"Analemma", a method which we could classify under descriptive
geometry. It existed already before Ptolemy since he criticizes the
unsystematic definitions of his predecessors. What follows is
taken from Ptolemy's very elegant treatment of the subject.
Related procedures are known from Vitruvius, the Roman
architect under Augustus, and from Heron, who wrote about 70
years before Ptolemy.

The problem itself concerns the theory of sun-dials in the
simplest form of a vertical "gnomon". Mathematically the problem
consists in defining proper spherical coordinates for the position
of the sun at a given moment for a given geographical latitude and
then to find graphically in a plane the arcs which represent these
coordinates. Ptolemy's predecessors—he calls them the "ancients"
without telling us who they were—operated with the following
system: consider, e. g., the octant of the celestial sphere which is
bounded by the planes of the horizon, of the meridian, and of
the vertical which is perpendicular to the two first mentioned
planes (cf. Fig. 47). The center of the sphere is the observer, the
vertical axis the gnomon. Let Σ be the position of the sun as seen
at the given moment. Then two planes are passed through Σ:
a vertical plane which contains the gnomon, and an inclined
plane which contains the axis OS. In this way three angles are

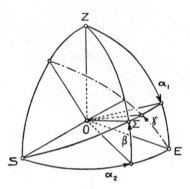

Fig. 47.

defined: α_1 and α_2 in the coordinate planes and β in the vertical plane through Σ. Each pair $\alpha_1\alpha_2$ or $\alpha_1\beta$ or $\alpha_2\beta$ can be used to define the position of Σ. To this "the ancients" added one more angle. For given geographical latitude φ the position of the equator plane is given; its intersection with the plane $S\Sigma$ defines a new angle γ and it is clear that also the pairs $\alpha_2\gamma$ and $\beta\gamma$ can be used to define the position of Σ since for given γ also the arc α_1 is fixed and vice versa.

Ptolemy would not tolerate such inelegant definitions. Using the same three orthogonal axes he would pass three planes through Σ and one of the axes, respectively. Then six angles $\alpha_1, \ldots, \beta_3$ are defined, as indicated in Fig. 48, two of which always suffice to define the position of Σ. The arrangement is strictly cyclical, all angles are acute and counted from one of the three orthogonal axes.

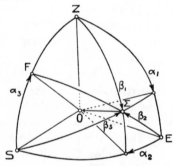

Fig. 48.

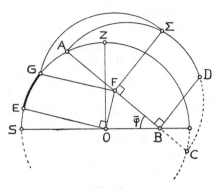

Fig. 49.

Ptolemy now proceeds to construct these six angles. As an illustration of the method followed, it will suffice to take the case of β_2 (called "hectemoros"). As plane of construction we use the plane SOZ of the meridian (Fig. 49). For given φ we know the angle $\bar{\varphi} = 90 - \varphi$ under which the equator is inclined to the horizon. For a given moment we also know the solar longitude and thus its declination; we also know for a given seasonal hour (here assumed to be an hour before noon) the fraction travelled by the sun between sunrise and noon. Thus we construct the trace ABC of the plane of the parallel-circle travelled at the given day and swing this plane into the plane of the meridian. If DB is perpendicular to ABC, then D represents the point of sunrise, A the point of culmination for the given day[1]). Thus we can find the position Σ of the sun for the given hour on the arc DA. Construct ΣF perpendicular to AB and make $FG = F\Sigma$. If E is placed such that EO is perpendicular to OF, then EG is the arc β_2 which we wanted to find. Indeed, E is the east-point of the horizon swung into our plane of construction and OF is the trace of the plane of the hectemoros in the meridian. Thus G is the place of the sun in the plane of the hectemoros which appears rotated about OF. Thus $EG = E\Sigma$.

Similar procedures lead to the determination of the other angles. It would lead us too far to present the details here but it is of great principal interest to mention the mechanization of these

---

[1]) Obviously DA/DC gives the ratio of daylight to night for the date in question.

constructions such that they can rapidly be carried out for any geographical latitude φ and for arbitrary solar longitude λ. All cases have some elements in common which do not depend on φ and λ, e. g. in Fig. 49 the circle of the meridian and the parallels to the equator. These circles are to be engraved on a plate of metal or stone or—in a cheaper model painted in black or red on a wooden disk. About the center is drawn the meridian and concentric with it a circle which indicates the angles φ corresponding to the seven "climates" which have longest daylight of 13^h $13\frac{1}{2}^h$ ... 16^h respectively. On the proper diameter are indicated points which correspond to the hourly position of the sun at equinox. The plate is then covered with wax so that additional lines which depend on the special values of λ and φ can be easily drawn. The disc can rotate about its center and a straight edge with right angle permits one to connect points of the different graduated circles corresponding to the swing of the projections about the proper axes. In this way it is possible to determine the angles in question by a procedure which is now called nomographical. In principle it is of the same character as the determination of angles on the celestial sphere by means of the circles and disks of an astrolabe. This is a good illustration of the fact that "Greek" mathematics was by no means rigidly restricted to some "classical" problems, as so many modern authors seem to believe.

88. In the notes to Chapter VI No. 62 (p. 181) I have given an example of the relationship between the theory of conic sections and "geometric algebra" as it existed in the time of Archimedes and Apollonius. This aspect does not exhaust by far the importance of the ancient study of the theory of conic sections. A large part of Apollonius's work on conic sections deals with problems which were later, in the 19th century, classified as projective or synthetic geometry—fields, which were developed in direct continuation of the ancient theory. Islamic and late medieval optics (Ibn al-Haitham, Kepler) are concerned with the focal properties. In antiquity the conic sections are needed for the theory of sundials and I have conjectured that the study of these curves originated from this very problem.

In another case the astronomical use of the theory of conic sections is almost certain, that is, the proof of the fact that stereographic projection maps circles on the sphere into circles in the plane. This fact is a consequence of a theorem, proved by Apollo-nius[1]), that there exist on every oblique circular cone two families of circular sections and it is easy to see that a circle on the sphere and its image projected from one pole of the equator onto its plane are exactly in the relation which Apollonius requires for elements of the two families. In the existing works no proof of this fact has come down to us but the circle-preserving quality of stereographic projection is commonly used in the treatises on the astrolabe and in Ptolemy's "Planisphaerium"[2]).

This work of Ptolemy is another example of the combination of descriptive geometry and trigonometric methods, and of practical devices which, by themselves, led to the instrument later known as the "astrolabe". The problem to be solved consists in the determination of the centers and radii of the circles which are the images of circles on the celestial sphere when projected from the south pole onto the plane of the equator. In order to determine these quantities the plane of the equator is used simultaneously as plane of mapping and as plane of construction orthogonal to it. For example, the radius of the circle which represents the ecliptic is found as follows (Fig. 50). The circle $abgd$ represents the equator, and a plane perpendicular to it in which the diameter bd is the axis with d as south pole. Now make $az = ng = gh =$

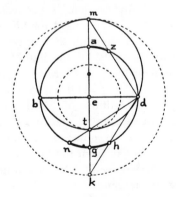

Fig. 50.

[1]) Conics I, 5.
[2]) Heath is incorrect when he says (Greek Mathematics II p. 292) that Ptolemy proves our theorem in special cases. The proofs, referred to by Heath, concern the fact that also the images of great circles intersect each other in diametrically opposite points but these proofs make use of the circularity of the images.

23;51° = ε the obliquity of the ecliptic and project the points
n, h, and z from d onto the diameter ag which now represents the
trace of the plane of the equator. Then tm is the diameter of the
ecliptic, b the vernal equinox, t and m the solstices through which
go the solstitial circles which remain tangential to the ecliptic in
all its possible positions.

In similar fashion all the celestial coordinate systems can be
represented as families of circles in the plane. In this way it is
possible to determine by means of plane geometry the rising times
of the zodiacal signs, both for sphaera recta ($\varphi = 0$) and for
general geographical latitudes. Since the method of stereographic
projection precedes in all probability the invention of spherical
trigonometry one sees here another way of finding the answer to
problems which later were solved directly from spherical triangles.

89. The problem of mapping the sphere onto a plane arises
once more in the field of geography. Again Ptolemy is the main
source of information for us. In the first book of his "Geography"
he gives the rules for constructing a grid of curves representing
the circles of constant geographical longitude and latitude respectively. The following will give a general impression of the
methods of the foundations of Greek cartography without making
any attempt to investigate the prehistory of geographical mapping.
We may only remark that Ptolemy's predecessor, Marinus of
Tyre, who wrote about 110 A.D., used for his map a rectangular
coordinate system in which the units that represented geographical
longitudes were $\frac{4}{5}$ of the units for the latitudes φ. Consequently
the spacing of the meridians is everywhere the same as at a
latitude for which cos $\varphi = \frac{4}{5}$. This is with sufficient accuracy
satisfied for $\varphi = 36$, the latitude of Rhodes. Thus the map of
Marinus preserves distances on all meridians and on the parallel
of Rhodes. Distances in all other directions are increasingly
distorted as one moves away from the latitude of Rhodes.

The mappings suggested by Ptolemy are much more sophisticated. Two belong to the general class of conic projections, the
third is a perspective representation of the terrestrial globe. I
shall give a short description of all three methods using modern
terminology. Ptolemy assumes that the inhabited part of the
earth, the "oikoumene", lies within 63° northern latitude (Thule)
and 16;25° southern latitude ("anti-Meroe", a parallel as far south

of the equator as Meroe in Nubia lies north of it). In longitude the oikoumene is assumed to extend 180°, and we shall count longitudes L from $-90°$ at the western limit to $+90$ at the eastern boundary. The first conic projection uses polar coordinates (Fig. 51) which we call r and δ. All meridians are mapped on radii, all parallels of latitudes on circles $r = $ const. Then three requirements are made: (a) no distortion of lengths on meridians, (b) nor on the parallel of Rhodes, and (c) the ratio of lengths on the parallel of Thule and on the equator should be preserved. The first condition implies that

$$r = \bar{\varphi} + c \qquad \bar{\varphi} = 90 - \varphi$$

(c a constant counted in degrees). The second condition means for the coordinates r_0, φ_0 of Rhodes

$$\frac{\pi}{180} r_0 \delta = L \cos \varphi_0$$

or

$$\delta = \frac{180 \cdot \cos \varphi_0}{\pi(\bar{\varphi}_0 + c)} \cdot L.$$

Finally the constant c can be determined from the last condition

$$\frac{\bar{\varphi}_1 + c}{90 + c} = \cos \varphi_1$$

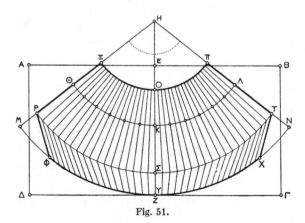

Fig. 51.

where $\varphi_1 = 63$ (latitude of Thule). This leads to the value $c = 25$ and thus to the values of $r = \bar{\varphi} + 25$ for each latitude[1]). Fig. 51 is drawn to scale and shows the resulting boundaries according to Ptolemy's construction; the arc $\varXi O\varPi$ represents $180°$ of the parallel of Thule, MN of the parallel of $-16;25$, $P\varSigma T$ of the equator; K lies on the parallel of Rhodes. In order to avoid the distortions in longitude on the southern boundary, Ptolemy arbitrarily changes the mapping south of the equator by dividing $\varPhi ZX$ in segments of a length as they would have had at the latitude 16;25 north of the equator.

The discontinuity in the direction of the meridians at the equator seemed to him less detrimental than the enlargement of the picture beyond the length of the equator. Here mathematical consistency was sacrificed to implausible appearance.

The second method of projection (Fig. 52) was devised to remedy this defect and to obtain a representation which is closer to the impression of gradually curving meridians. Again Ptolemy requires that the radial distances correctly reflect latitudinal differences though the radii no longer represent meridians (except for the central meridian $L = 0$). Thus we have as before

(1) $$r = \bar{\varphi} + c.$$

For the circular arcs which now represent the meridians we determine three points by the following conditions: preservation of length on the parallel of Thule ($\varphi_1 = 63$), on the parallel of Syene which lies on the Tropic of Cancer ($\varphi_2 = \varepsilon = 23;50$), and on the parallel of Anti-Meroe $\varphi_3 = -16;25$. Consequently we have

(2) $$\frac{\pi}{180} r_i \delta = L \cos \varphi_i \qquad i = 1, 2, 3.$$

The value of c in (1) determines the curvature of the limiting parallels of the map. Ptolemy chooses $c = 180$ on the basis of a simple geometrical consideration by means of which he obtains

[1]) Consequently the north pole is mapped on the circle $r = c = 25$ (dotted in Fig. 51).

for the map of the oikoumene dimensions reasonably like the actual ratios.

With c fixed, the map can now be constructed (Fig. 52). H is the common center of the parallels of latitude, E lies on the parallel of Syene and therefore HE is to be made $180-23;50 = 156;10$, and $EO = 23;50 + 16;25 = 40;15$ determines the southern boundary. $HN = 180 - 90 = 90$ gives the image N of the

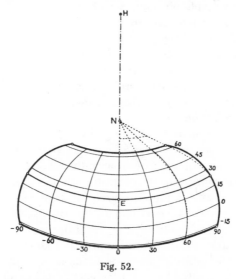

Fig. 52.

north pole. Substituting for L the same value (e. g. $L = 90$ for the eastern boundary) in all three equations resulting from the use of φ_1, φ_2, φ_3 respectively gives three points through which the meridian of longitude L must pass. In fact the curves $L = \text{const}$ are transcendental curves but Ptolemy replaces them by the circle arc which is determined by the three points (L, φ_i).

It may seem that this last approximation is a very crude one, though convenient for the actual construction of the grid. In fact, however, it is a remarkably good approximation, within the area $-16;25 \leqq \varphi \leqq 63$ and $-90 \leqq L \leqq 90$ which contains the oikoumene. If (2) were required for all values of φ, one would obtain the so-called "Bonne-projection" which preserves length on all parallels of latitude. In Fig. 52. I have added in dotted lines meridians of the Bonne projection; only for the

extreme-north-eastern area the deviation between the meridians
of Bonne and Ptolemy reach visible proportions.

The cartographic designs discussed so far are mappings in the
modern mathematical sense of this word: mathematical relations
are defined which relate a point with coordinates L, φ of the
sphere to a point with coordinates r, δ in the plane. This relationship is not obtainable, however, as the image of the sphere seen
by an eye in suitable position. A truly perspective picture occurs,
however, in a presentation of the terrestrial globe in Book VII
Ch. 6 of Ptolemy's "Geography" though in a very inconsistent
combination with a mapping of the Bonne type. Ptolemy assumes
a terrestrial globe mounted between rings which represent the
arctic circles, the solstitial circles and equator and ecliptic. A
perspective picture of these rings is then constructed, seen from
a center of projection located in such a fashion that no part of the
area of the oikoumene is obscured by a ring. The globe within
the rings, however, is not represented in perspective but simply
as a circular frame of a map similar to the second above-described
networks. It is more a book illustration than a real map which is
described here, and is the only case in all of Ptolemy's writings
where he displays an inconsistent and totally useless construction,
thus foreshadowing the taste of the Middle Ages.

90. The history of mathematics provides good illustrations for
the fact that continuity of tradition alone is not sufficient to keep
a scientific field alive. The "Elements" of Euclid formed for
centuries the basis of mathematical instruction. Nevertheless the
significance of its axiomatic structure was not understood until
the problems connected with the foundations of analysis led the
mathematicians of the 19th century to similar methods. A particular case in point is the theory of proportions in Book V which was
only seen in its proper role within Greek mathematics through
the theory of irrational numbers and continuity developed by
R. Dedekind since 1858. Similarly it required the recent development of formal logic to discover the existence of similar systems
in the writings of Aristotle and the Stoic and Megaric philosophers.

I am not competent to discuss the modern aspects of Aristotle's
logic and of the later schools. One aspect, however, should be
mentioned which concerns mathematics in its narrower sense,
that is the use of letters in the formulation of syllogisms. For

example the figure later on called "Barbara" is given as "If A is predicated of all B and B is predicated of all C, then A must be predicated of all C[1])". The introduction of variables, represented by letters into logic would seem to have constituted the essential step to a formulation of mathematical rules which we would call "Algebra". Such a development would have been still more natural since the axiomatic approach to mathematics originated in the same time and among the same circle of men. Yet nothing of this type happened and the origin of algebra is totally independent of the existence of an algebraic notation in one of the most famous philosophical works of antiquity.

This is a good illustration for the futility of any attempt to reconstruct "reasons" for the incidents of historical events. Similarly the absence of algebraic notations should not have prevented the Greek geometers from developing what was called in the 19th century "synthetic" and "projective" geometry since many of the basic concepts were ready at hand in the works of Apollonius. Again such a "natural" development did not take place and all that we may ever hope to establish in historical research is facts and conditions but never causes.

BIBLIOGRAPHY TO APPENDIX II

There is no lack of histories of Greek mathematics and every library catalogue will suffice to identify many of them. I therefore restrict myself here to quoting a few comparatively recent publications which have not yet become commonly used.

A work of Archimedes on the regular heptagon has been recovered through an Arabic translation, a German summary of which was published by C. Schoy in his work "Die trigonometrischen Lehren des persischen Astronomen ... Al-Bīrūnī" (Hannover 1927) p. 74–91.

The "Sphaerica" of Menelaos, likewise only preserved in Arabic translation, or version, by Abū Naṣr Manṣūr (about 1000 A.D.) is available in a critical edition with German translation by Max Krause (Abhandlungen der Gesellschaft der Wissenschaften zu Göttingen, philol.-hist. Kl., 3 Folge, Nr. 17, Berlin, Weidmann, 1936).

A new edition of one of the earliest Greek mathematical works that have come down to us (written perhaps about 330–300 B.C.) was given by J. Mogenet "*Autolycus de Pitane*; histoire du text, suivie de l'édition critique des traités de la Sphère en Mouvement et des Levers et Couchers (Louvain 1950,

[1]) Translation of Analytica priora I, 4 25 ᵇ 37 by Lukasiewicz (Aristotle's Syllogistic p. 3 and p. 10).

Recueil de Travaux d'Histoire et de Philologie, 3ᵉ sér., fasc. 37). In the meantime
O. Schmidt made the interesting discovery that the two "books" of the "On
Risings and Settings" are actually only two versions of the same work; cf. the
article "Some critical remarks about Autolycus' On Risings and Settings" in
the transactions of Den 11te skandinaviske Matematikerkongress i Trondheim
22–25 August 1949 (published Oslo 1952) p. 202–209.

A. Lejeune, *Euclide et Ptolemée*; deux stades de l'optique géométrique
grecque (Louvain 1948, Recueil de Travaux d'Histoire et de Philologie, 3ᵉ sér.,
fasc. 31) is a work of great historical and methodological interest. In a careful
analysis we see here the progress from a strictly geometrical optics to a theory
of binocular vision and physiological optics based on empirical data and
systematic experiments. Ptolemy's optical theories touch also upon the problem
of the three-dimensionality of space, a subject on which he also wrote a special
treatise, now lost.[1]) Ptolemy's *Optics* is only preserved in a Latin version of an
Arabic translation, (edited by A. Lejeune, Louvain 1956, l. c., 4ᵉ sér., fasc. 8).

NOTES AND REFERENCES TO APPENDIX II

ad 85. The geodetical problem, mentioned on p. 210, is known as the
Snellius-Pothenot problem. It was solved by W. Snellius in his "Eratosthenes
Batavus" Leiden 1617 p. 203 f. J. A. C. Oudemans stated in Vierteljahrschrift
der astronomischen Gesellschaft 22 (1887) p. 345 that Ptolemy's problem was
identical with the problem of Snellius. This, however, is not the case. Ptolemy
assumes as given δ_1, δ_2, and α_1, α_2 and R whereas Snellius knows beyond δ_1
and δ_2 all three sides s_1, s_2, s_3 of the triangle.

The problem has also been discussed by Delambre, Hist. Astron. Ancienne II
p. 164 ff. Cf. also Tropfke, Geschichte der Elementarmathematik V, 2nd ed.
1923 p. 97.

ad 87. Literature concerning the "Analemma": Ptolemy's treatise is published
by Heiberg in Ptolemaeus, Opera II, p. 189–223 (1907). For the discussion of
the method, cf. Delambre, Histoire de l'astronomie ancienne II p. 458–471
(Paris 1817); J. Drecker, Theorie der Sonnenuhren (in: Bassermann-Jordan,
Geschichte der Zeitmessung und der Uhren, Bd. I, E; Berlin 1925); and in
particular the article by P. Luckey, Das Analemma von Ptolemäus, Astron.
Nachrichten 230 No. 5498 p. 17–46 (1927), to whom we owe the understanding
of the nomographic procedures of Ptolemy's "Analemma".

For Vitruvius and Heron cf. O. Neugebauer, Über eine Methode zur Distanz-
bestimmung Alexandria-Rom, Kgl. Danske Vidensk. Selsk., hist.-filol. medd.
26, 2 and 26, 7 (1938–1939).

ad 88. O. Neugebauer, On the astronomical origin of the theory of conic
sections. Proc. Amer. Philos. Soc. 92 (1948) p. 136–138. My argument is based
on the fact that the earliest form of the theory assumes that always one generating
line is perpendicular to the intersecting plane, an arrangement as in the case
of the gnomon with respect to the plane upon which the shadow is cast. The
difficulty of this conjecture lies in the fact that no sun-dials seem to be preserved
in which the gnomon is directed toward the culminating sun.

In passing, it may be remarked that the theory of sun-dials is perhaps the

_____
[1]) Cf. Ptolemaeus, Opera astron. minora, ed. Heiberg, p. 265 f.

origin of one of the "classical" problems of Greek mathematics, namely, the trisection of angles. We know from Pappus (Collection IV, 27) that he used the conchoid of Nicomedes (2nd century B.C.) in order to trisect an angle in connection with a work of Diodorus (first century B.C.) on the theory of sun dials. Here the problem arises of constructing the 12th part of the arc which the sun is above the horizon, because this is the equivalent of one "seasonal hour".

ad 89. The best discussion of Ptolemy's theory of map projection is given by H. v. Mžik–F. Hopfner, Des Klaudios Ptolemaios Einführung in die darstellende Erdkunde, Klotho 5 (1938). Cf. Also H. Berger, Geschichte der wissenschaftlichen Erdkunde der Griechen, 2. Aufl., Leipzig 1903 (p. 632 ff.).

A great variety of ancient maps is reproduced in the monumental work Claudii Ptolemaei Geographiae Codex Urbinas Graecus 82 (= Codices e Vaticanis selecti vol. 19), Leiden-Leipzig 1932 (4 vols.), edited and commented by J. Fischer. It must be emphasized, however, that the "Geography" in eight books, as it exists today, is in all probability not the work of Ptolemy but rather a Byzantine compilation, as L. Bagrow has shown ("The Origin of Ptolemy's Geographia" Geografiska Annaler 1943 p. 318–387, in particular p. 368 ff.). The mathematical sections, ("Book I") are unquestionably genuine.

Ptolemy himself was fully aware of the fact that his second conic projection, with circular arcs as meridians, was only an approximation, though a very good one, to a mapping in which distances were preserved on all parallels of latitude. It is only for the simplicity of construction that he restricted himself to three parallels. The accurate meridian lines, for the case of the north pole as center, were given by Johannes Werner in connection with his translation of the first book of Ptolemy's Geography (Nürnberg 1514 and again Ingolstadt 1533 with an introduction by Petrus Apianus). This method of projection became very popular after it was used in an atlas by R. Bonne (1787) and was later adopted for the mapping of France on the recommendation of Laplace.

The fact that this "Bonne-projection" is area-preserving was, of course, unknown in antiquity.

ad 90. For the relationship between the Greek theory of irrationals and its modern counterpart cf. R. Dedekind's monographs "Stetigkeit und irrationale Zahlen" (1872) and "Was sind und was sollen die Zahlen" (1888) as well as his correspondence with R. Lipschitz (R. Dedekind, Gesammelte mathematische Werke III p. 469–479). Cf. furthermore: O. Becker, Eudoxos-Studien, Quellen und Studien zur Geschichte der Mathematik, Abt. B vol. 2, p. 311–333; p. 369–387; vol. 3, p. 236–244; p. 370–388; p. 389–410 (1932–1934).

K. v. Fritz, The discovery of incommensurability by Hippasus of Metapontum, Annals of Mathematics 46 (1945) p. 242–264.

Van der Waerden, Die Arithmetik der Pythagoreer, Mathem. Annalen 120 (1947/1949) p. 127–153; p. 676–700.

In recent years a large number of books and monographs have been written on problems of ancient formal logic. It may suffice to quote the following works:

Jan Lukasiewicz, Aristotle's Syllogistic from the standpoint of modern formal logic, Oxford, Clarendon Press, 1951.

I. M. Bocheński, Ancient Formal Logic, North-Holland Publishing Company, Amsterdam 1951.

Benson Mates, Stoic Logic, Univ. of California Press, 1953.

228

THE ZODIACAL SIGNS

♈ Aries   ♎ Libra
♉ Taurus   ♏ Scorpio
♊ Gemini   ♐ Sagittarius
♋ Cancer   ♑ Capricorn
♌ Leo   ♒ Aquarius
♍ Virgo   ♓ Pisces

THE PLANETARY SYMBOLS

♄ Saturn   ☉ Sun
♃ Jupiter   ♀ Venus
♂ Mars   ☿ Mercury

CHRONOLOGICAL TABLE

Cf. also the Frontispiece.
Dates are only approximate.

−1700	Old Babylonian	360	Theon Alex.
−1300	Seti I	380	Paulus Alex.
− 650	Ashurbanipal	450	Proclus
− 430	Meton	500	Āryabhaṭa
− 375	Archytas	500	Rhetorios
− 370	Eudoxus	550	Varāha Mihira
− 350	Aristotle	650	Severus Sebokht
− 311	beg. of Seleucid Era	650	Brahmagupta
− 300	Euclid	825	al-Khwārizmī
− 275	Aristarch	850	Abū Maʿshar
− 275	Aratus	900	al-Battānī
− 275	Berossos	1000	ibn-Yūnus
− 250	Eratosthenes	1000	ibn al-Haitham
− 240	Archimedes	1000	al-Bīrūnī
− 200	Apollonius	1000	Suidas
− 150	Hipparchus	1130	Adelard of Bath
− 100	Theodosius(?)	1150	Bhascara
− 100	Teukros(?)	1170	Maimonides
− 75	Geminus	1250	Alfonso X
+ 10	Manilius	1250	Bar Hebraeus
60	Pliny	1430	Ulūgh Beg
75	Heron	1500	Copernicus
75	latest cuneif. text	1540	Rheticus
100	Menelaos	1575	Tycho Brahe
150	Ptolemy	1600	Kepler
160	Vettius Valens	1680	Halley
160	Galen	1686	Newton
340	Pappus	1700	Cassini
350	Diophantus(??)	1760	LeGentil

PLATE 1

"September", Book of Hours of the Duke of Berry.

PLATE 2

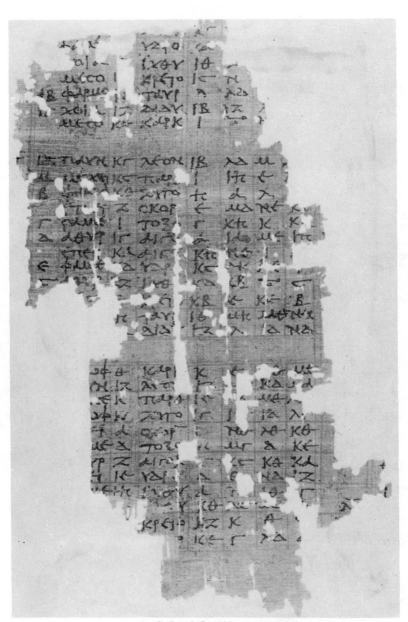

P. Lund, Inv. 35 a.

PLATE 3

BM 85194 Obv.

PLATE 4

b. VAT 12770.

a. VAT 7858.

PLATE 5

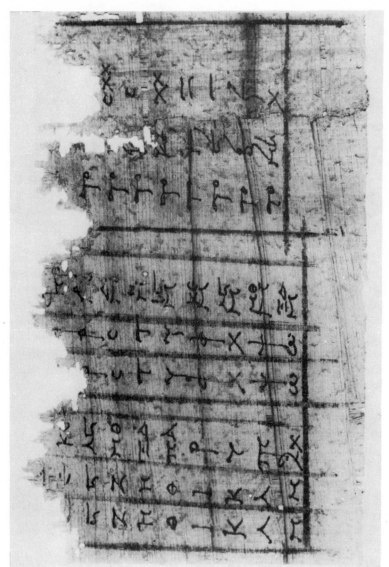

PLATE 4

b. VAT 12770.

a. VAT 7858.

PLATE 5

P. Cairo, Inv. 65445.

PLATE 6

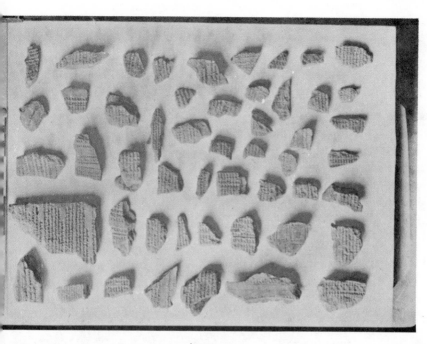

b. Warka Fragments.

a. YBC 7289.

PLATE 7

a. Plimpton 322.

b. A 3412 Rev.

PLATE 8

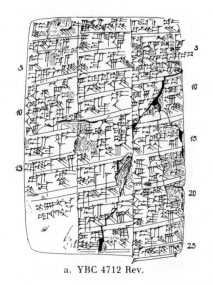

a. YBC 4712 Rev.

b. Sp. II, 62 = BM 34589.

PLATE 9

a. YBC 4712 Rev.

b. Sp. II, 62 = BM 34589.

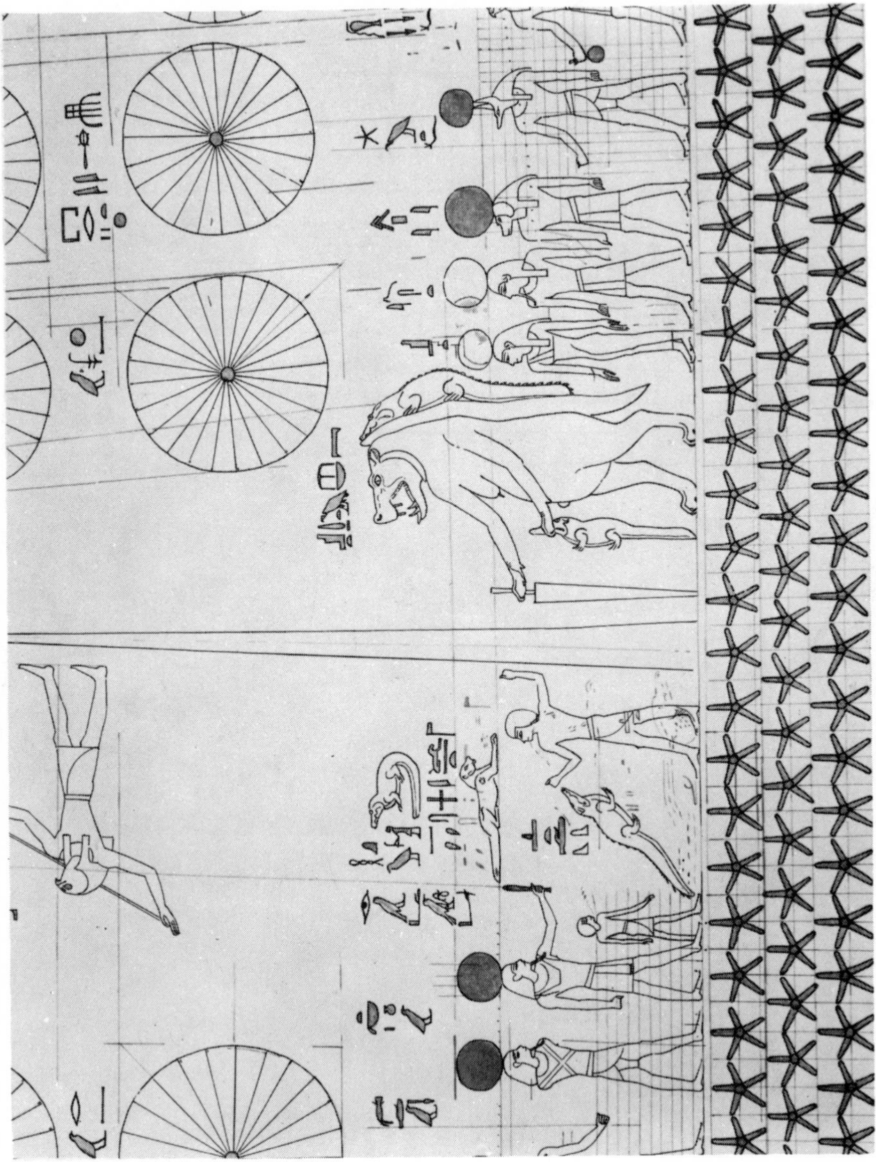

PLATE 10

From the Ceiling in the Tomb of Senmut.

PLATE 11

From the Tomb of Ramses VII.

PLATE 12

P. Cornell, Inv. 69.

PLATE 13

Demotic Planetary Tables, Liverpool.

PLATE 14

690.

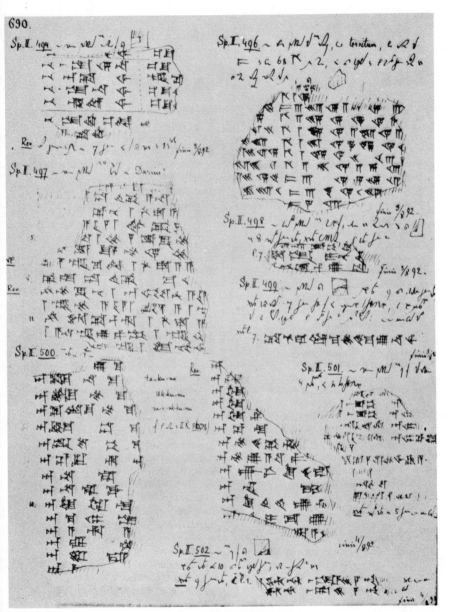

From Strassmaier's Notebook.

INDEX

236 Index

A CATALOG OF SELECTED DOVER
BOOKS IN ALL FIELDS OF INTEREST

DRAWINGS OF REMBRANDT, edited by Seymour Slive. Updated Lippmann, Hofstede de Groot edition, with definitive scholarly apparatus. All portraits, biblical sketches, landscapes, nudes. Oriental figures, classical studies, together with selection of work by followers. 550 illustrations. Total of 630pp. 9⅛ × 12¼.
21485-0, 21486-9 Pa., Two-vol. set $25.00

GHOST AND HORROR STORIES OF AMBROSE BIERCE, Ambrose Bierce. 24 tales vividly imagined, strangely prophetic, and decades ahead of their time in technical skill: "The Damned Thing," "An Inhabitant of Carcosa," "The Eyes of the Panther," "Moxon's Master," and 20 more. 199pp. 5⅜ × 8½. 20767-6 Pa. $3.95

ETHICAL WRITINGS OF MAIMONIDES, Maimonides. Most significant ethical works of great medieval sage, newly translated for utmost precision, readability. Laws Concerning Character Traits, Eight Chapters, more. 192pp. 5⅜ × 8½.
24522-5 Pa. $4.50

THE EXPLORATION OF THE COLORADO RIVER AND ITS CANYONS, J. W. Powell. Full text of Powell's 1,000-mile expedition down the fabled Colorado in 1869. Superb account of terrain, geology, vegetation, Indians, famine, mutiny, treacherous rapids, mighty canyons, during exploration of last unknown part of continental U.S. 400pp. 5⅜ × 8½. 20094-9 Pa. $6.95

HISTORY OF PHILOSOPHY, Julián Marías. Clearest one-volume history on the market. Every major philosopher and dozens of others, to Existentialism and later. 505pp. 5⅜ × 8½. 21739-6 Pa. $8.50

ALL ABOUT LIGHTNING, Martin A. Uman. Highly readable non-technical survey of nature and causes of ~~ing, th ~~rstorms, ball lightning, St. Elmo's Fire, much more. Illustrated. 192pp. ⌐. 25237-X Pa. $5.95

SAILING ALONE AROUND THE WORLD, Cap~ ~hua Slocum. First man to sail around the world, alone, in small boat. One of great ⌐ ~eamanship told in delightful manner. 67 illustrations. 294pp. 5⅜ × 8½. 326-3 Pa. $4.50

LETTERS AND NOTES ON THE MANNERS, CUSTOMS AND CONDI-TIONS OF THE NORTH AMERICAN INDIANS, George Catlin. Classic account of life among Plains Indians: ceremonies, hunt, warfare, etc. 312 plates. 572pp. of text. 6⅛ × 9¼. 22118-0, 22119-9 Pa. Two-vol. set $15.90

ALASKA: The Harriman Expedition, 1899, John Burroughs, John Muir, et al. Informative, engrossing accounts of two-month, 9,000-mile expedition. Native peoples, wildlife, forests, geography, salmon industry, glaciers, more. Profusely illustrated. 240 black-and-white line drawings. 124 black-and-white photographs. 3 maps. Index. 576pp. 5⅜ × 8½. 25109-8 Pa. $11.95

THE BOOK OF BEASTS: Being a Translation from a Latin Bestiary of the Twelfth Century, T. H. White. Wonderful catalog real and fanciful beasts: manticore, griffin, phoenix, amphivius, jaculus, many more. White's witty erudite commentary on scientific, historical aspects. Fascinating glimpse of medieval mind. Illustrated. 296pp. 5⅜ × 8¼. (Available in U.S. only)   24609-4 Pa. $5.95

FRANK LLOYD WRIGHT: ARCHITECTURE AND NATURE With 160 Illustrations, Donald Hoffmann. Profusely illustrated study of influence of nature—especially prairie—on Wright's designs for Fallingwater, Robie House, Guggenheim Museum, other masterpieces. 96pp. 9¼ × 10¾.   25098-9 Pa. $7.95

FRANK LLOYD WRIGHT'S FALLINGWATER, Donald Hoffmann. Wright's famous waterfall house: planning and construction of organic idea. History of site, owners, Wright's personal involvement. Photographs of various stages of building. Preface by Edgar Kaufmann, Jr. 100 illustrations. 112pp. 9¼ × 10.
23671-4 Pa. $7.95

YEARS WITH FRANK LLOYD WRIGHT: Apprentice to Genius, Edgar Tafel. Insightful memoir by a former apprentice presents a revealing portrait of Wright the man, the inspired teacher, the greatest American architect. 372 black-and-white illustrations. Preface. Index. vi + 228pp. 8¼ × 11.   24801-1 Pa. $9.95

THE STORY OF KING ARTHUR AND HIS KNIGHTS, Howard Pyle. Enchanting version of King Arthur fable has delighted generations with imaginative narratives of exciting adventures and unforgettable illustrations by the author. 41 illustrations. xviii + 313pp. 6⅛ × 9¼.   21445-1 Pa. $5.95

THE GODS OF THE EGYPTIANS, E. A. Wallis Budge. Thorough coverage of numerous gods of ancient Egypt by foremost Egyptologist. Information on evolution of cults, rites and gods; the cult of Osiris; the Book of the Dead and its rites; the sacred animals and birds; Heaven and Hell; and more. 956pp. 6⅛ × 9¼.
22055-9, 22056-7 Pa., Two-vol. set $20.00

A THEOLOGICO-POLITICAL TREATISE, Benedict Spinoza. Also contains unfinished *Political Treatise*. Great classic on religious liberty, theory of government on common consent. R. Elwes translation. Total of 421pp. 5⅜ × 8½.
20249-6 Pa. $6.95

INCIDENTS OF TRAVEL IN CENTRAL AMERICA, CHIAPAS, AND YUCATAN, John L. Stephens. Almost single-handed discovery of Maya culture; exploration of ruined cities, monuments, temples; customs of Indians. 115 drawings. 892pp. 5⅜ × 8½.   22404-X, 22405-8 Pa., Two-vol. set $15.90

LOS CAPRICHOS, Francisco Goya. 80 plates of wild, grotesque monsters and caricatures. Prado manuscript included. 183pp. 6⅜ × 9⅜.   22384-1 Pa. $4.95

AUTOBIOGRAPHY: The Story of My Experiments with Truth, Mohandas K. Gandhi. Not hagiography, but Gandhi in his own words. Boyhood, legal studies, purification, the growth of the Satyagraha (nonviolent protest) movement. Critical, inspiring work of the man who freed India. 480pp. 5⅜ × 8½. (Available in U.S. only)
24593-4 Pa. $6.95

ILLUSTRATED DICTIONARY OF HISTORIC ARCHITECTURE, edited by Cyril M. Harris. Extraordinary compendium of clear, concise definitions for over 5,000 important architectural terms complemented by over 2,000 line drawings. Covers full spectrum of architecture from ancient ruins to 20th-century Modernism. Preface. 592pp. 7½ × 9⅝. 24444-X Pa. $14.95

THE NIGHT BEFORE CHRISTMAS, Clement Moore. Full text, and woodcuts from original 1848 book. Also critical, historical material. 19 illustrations. 40pp. 4⅝ × 6. 22797-9 Pa. $2.25

THE LESSON OF JAPANESE ARCHITECTURE: 165 Photographs, Jiro Harada. Memorable gallery of 165 photographs taken in the 1930's of exquisite Japanese homes of the well-to-do and historic buildings. 13 line diagrams. 192pp. 8⅜ × 11¼. 24778-3 Pa. $8.95

THE AUTOBIOGRAPHY OF CHARLES DARWIN AND SELECTED LETTERS, edited by Francis Darwin. The fascinating life of eccentric genius composed of an intimate memoir by Darwin (intended for his children); commentary by his son, Francis; hundreds of fragments from notebooks, journals, papers; and letters to and from Lyell, Hooker, Huxley, Wallace and Henslow. xi + 365pp. 5⅜ × 8. 20479-0 Pa. $5.95

WONDERS OF THE SKY: Observing Rainbows, Comets, Eclipses, the Stars and Other Phenomena, Fred Schaaf. Charming, easy-to-read poetic guide to all manner of celestial events visible to the naked eye. Mock suns, glories, Belt of Venus, more. Illustrated. 299pp. 5¼ × 8¼. 24402-4 Pa. $7.95

BURNHAM'S CELESTIAL HANDBOOK, Robert Burnham, Jr. Thorough guide to the stars beyond our solar system. Exhaustive treatment. Alphabetical by constellation: Andromeda to Cetus in Vol. 1; Chamaeleon to Orion in Vol. 2; and Pavo to Vulpecula in Vol. 3. Hundreds of illustrations. Index in Vol. 3. 2,000pp. 6⅛ × 9¼. 23567-X, 23568-8, 23673-0 Pa., Three-vol. set $36.85

STAR NAMES: Their Lore and Meaning, Richard Hinckley Allen. Fascinating history of names various cultures have given to constellations and literary and folkloristic uses that have been made of stars. Indexes to subjects. Arabic and Greek names. Biblical references. Bibliography. 563pp. 5⅜ × 8½. 21079-0 Pa. $7.95

THIRTY YEARS THAT SHOOK PHYSICS: The Story of Quantum Theory, George Gamow. Lucid, accessible introduction to influential theory of energy and matter. Careful explanations of Dirac's anti-particles, Bohr's model of the atom, much more. 12 plates. Numerous drawings. 240pp. 5⅜ × 8½. 24895-X Pa. $4.95

CHINESE DOMESTIC FURNITURE IN PHOTOGRAPHS AND MEASURED DRAWINGS, Gustav Ecke. A rare volume, now affordably priced for antique collectors, furniture buffs and art historians. Detailed review of styles ranging from early Shang to late Ming. Unabridged republication. 161 black-and-white drawings, photos. Total of 224pp. 8⅜ × 11¼. (Available in U.S. only) 25171-3 Pa. $12.95

VINCENT VAN GOGH: A Biography, Julius Meier-Graefe. Dynamic, penetrating study of artist's life, relationship with brother, Theo, painting techniques, travels, more. Readable, engrossing. 160pp. 5⅜ × 8½. (Available in U.S. only) 25253-1 Pa. $3.95

CATALOG OF DOVER BOOKS

HOW TO WRITE, Gertrude Stein. Gertrude Stein claimed anyone could understand her unconventional writing—here are clues to help. Fascinating improvisations, language experiments, explanations illuminate Stein's craft and the art of writing. Total of 414pp. 4⅜ × 6⅜. 23144-5 Pa. $5.95

ADVENTURES AT SEA IN THE GREAT AGE OF SAIL: Five Firsthand Narratives, edited by Elliot Snow. Rare true accounts of exploration, whaling, shipwreck, fierce natives, trade, shipboard life, more. 33 illustrations. Introduction. 353pp. 5⅜ × 8½. 25177-2 Pa. $7.95

THE HERBAL OR GENERAL HISTORY OF PLANTS, John Gerard. Classic descriptions of about 2,850 plants—with over 2,700 illustrations—includes Latin and English names, physical descriptions, varieties, time and place of growth, more. 2,706 illustrations. xlv + 1,678pp. 8½ × 12¼. 23147-X Cloth. $75.00

DOROTHY AND THE WIZARD IN OZ, L. Frank Baum. Dorothy and the Wizard visit the center of the Earth, where people are vegetables, glass houses grow and Oz characters reappear. Classic sequel to *Wizard of Oz*. 256pp. 5⅜ × 8. 24714-7 Pa. $4.95

SONGS OF EXPERIENCE: Facsimile Reproduction with 26 Plates in Full Color, William Blake. This facsimile of Blake's original "Illuminated Book" reproduces 26 full-color plates from a rare 1826 edition. Includes "The Tyger," "London," "Holy Thursday," and other immortal poems. 26 color plates. Printed text of poems. 48pp. 5¼ × 7. 24636-1 Pa. $3.50

SONGS OF INNOCENCE, William Blake. The first and most popular of Blake's famous "Illuminated Books," in a facsimile edition reproducing all 31 brightly colored plates. Additional printed text of each poem. 64pp. 5¼ × 7. 22764-2 Pa. $3.50

PRECIOUS STONES, Max Bauer. Classic, thorough study of diamonds, rubies, emeralds, garnets, etc.: physical character, occurrence, properties, use, similar topics. 20 plates, 8 in color. 94 figures. 659pp. 6⅛ × 9¼. 21910-0, 21911-9 Pa., Two-vol. set $14.90

ENCYCLOPEDIA OF VICTORIAN NEEDLEWORK, S. F. A. Caulfeild and Blanche Saward. Full, precise descriptions of stitches, techniques for dozens of needlecrafts—most exhaustive reference of its kind. Over 800 figures. Total of 679pp. 8⅛ × 11. Two volumes. Vol. 1 22800-2 Pa. $10.95
Vol. 2 22801-0 Pa. $10.95

THE MARVELOUS LAND OF OZ, L. Frank Baum. Second Oz book, the Scarecrow and Tin Woodman are back with hero named Tip, Oz magic. 136 illustrations. 287pp. 5⅜ × 8½. 20692-0 Pa. $5.95

WILD FOWL DECOYS, Joel Barber. Basic book on the subject, by foremost authority and collector. Reveals history of decoy making and rigging, place in American culture, different kinds of decoys, how to make them, and how to use them. 140 plates. 156pp. 7⅞ × 10¾. 20011-6 Pa. $7.95

HISTORY OF LACE, Mrs. Bury Palliser. Definitive, profusely illustrated chronicle of lace from earliest times to late 19th century. Laces of Italy, Greece, England, France, Belgium, etc. Landmark of needlework scholarship. 266 illustrations. 672pp. 6⅛ × 9¼. 24742-2 Pa. $14.95

ILLUSTRATED GUIDE TO SHAKER FURNITURE, Robert Meader. All furniture and appurtenances, with much on unknown local styles. 235 photos. 146pp. 9 × 12.   22819-3 Pa. $7.95

WHALE SHIPS AND WHALING: A Pictorial Survey, George Francis Dow. Over 200 vintage engravings, drawings, photographs of barks, brigs, cutters, other vessels. Also harpoons, lances, whaling guns, many other artifacts. Comprehensive text by foremost authority. 207 black-and-white illustrations. 288pp. 6 × 9.
24808-9 Pa. $8.95

THE BERTRAMS, Anthony Trollope. Powerful portrayal of blind self-will and thwarted ambition includes one of Trollope's most heartrending love stories. 497pp. 5⅜ × 8½.   25119-5 Pa. $8.95

ADVENTURES WITH A HAND LENS, Richard Headstrom. Clearly written guide to observing and studying flowers and grasses, fish scales, moth and insect wings, egg cases, buds, feathers, seeds, leaf scars, moss, molds, ferns, common crystals, etc.—all with an ordinary, inexpensive magnifying glass. 209 exact line drawings aid in your discoveries. 220pp. 5⅜ × 8½.   23330-8 Pa. $3.95

RODIN ON ART AND ARTISTS, Auguste Rodin. Great sculptor's candid, wide-ranging comments on meaning of art; great artists; relation of sculpture to poetry, painting, music; philosophy of life, more. 76 superb black-and-white illustrations of Rodin's sculpture, drawings and prints. 119pp. 8⅜ × 11¼.   24487-3 Pa. $6.95

FIFTY CLASSIC FRENCH FILMS, 1912–1982: A Pictorial Record, Anthony Slide. Memorable stills from Grand Illusion, Beauty and the Beast, Hiroshima, Mon Amour, many more. Credits, plot synopses, reviews, etc. 160pp. 8¼ × 11.
25256-6 Pa. $11.95

THE PRINCIPLES OF PSYCHOLOGY, William James. Famous long course complete, unabridged. Stream of thought, time perception, memory, experimental methods; great work decades ahead of its time. 94 figures. 1,391pp. 5⅜ × 8½.
20381-6, 20382-4 Pa., Two-vol. set $19.90

BODIES IN A BOOKSHOP, R. T. Campbell. Challenging mystery of blackmail and murder with ingenious plot and superbly drawn characters. In the best tradition of British suspense fiction. 192pp. 5⅜ × 8½.   24720-1 Pa. $3.95

CALLAS: PORTRAIT OF A PRIMA DONNA, George Jellinek. Renowned commentator on the musical scene chronicles incredible career and life of the most controversial, fascinating, influential operatic personality of our time. 64 black-and-white photographs. 416pp. 5⅜ × 8¼.   25047-4 Pa. $7.95

GEOMETRY, RELATIVITY AND THE FOURTH DIMENSION, Rudolph Rucker. Exposition of fourth dimension, concepts of relativity as Flatland characters continue adventures. Popular, easily followed yet accurate, profound. 141 illustrations. 133pp. 5⅜ × 8½.   23400-2 Pa. $3.50

HOUSEHOLD STORIES BY THE BROTHERS GRIMM, with pictures by Walter Crane. 53 classic stories—Rumpelstiltskin, Rapunzel, Hansel and Gretel, the Fisherman and his Wife, Snow White, Tom Thumb, Sleeping Beauty, Cinderella, and so much more—lavishly illustrated with original 19th century drawings. 114 illustrations. x + 269pp. 5⅜ × 8½.   21080-4 Pa. $4.50

SUNDIALS, Albert Waugh. Far and away the best, most thorough coverage of ideas, mathematics concerned, types, construction, adjusting anywhere. Over 100 illustrations. 230pp. 5⅜ × 8½. 22947-5 Pa. $4.00

PICTURE HISTORY OF THE NORMANDIE: With 190 Illustrations, Frank O. Braynard. Full story of legendary French ocean liner: Art Deco interiors, design innovations, furnishings, celebrities, maiden voyage, tragic fire, much more. Extensive text. 144pp. 8⅜ × 11¾. 25257-4 Pa. $9.95

THE FIRST AMERICAN COOKBOOK: A Facsimile of "American Cookery," 1796, Amelia Simmons. Facsimile of the first American-written cookbook published in the United States contains authentic recipes for colonial favorites—pumpkin pudding, winter squash pudding, spruce beer, Indian slapjacks, and more. Introductory Essay and Glossary of colonial cooking terms. 80pp. 5⅜ × 8½. 24710-4 Pa. $3.50

101 PUZZLES IN THOUGHT AND LOGIC, C. R. Wylie, Jr. Solve murders and robberies, find out which fishermen are liars, how a blind man could possibly identify a color—purely by your own reasoning! 107pp. 5⅜ × 8½. 20367-0 Pa. $2.00

THE BOOK OF WORLD-FAMOUS MUSIC—CLASSICAL, POPULAR AND FOLK, James J. Fuld. Revised and enlarged republication of landmark work in musico-bibliography. Full information about nearly 1,000 songs and compositions including first lines of music and lyrics. New supplement. Index. 800pp. 5⅜ × 8¼. 24857-7 Pa. $14.95

ANTHROPOLOGY AND MODERN LIFE, Franz Boas. Great anthropologist's classic treatise on race and culture. Introduction by Ruth Bunzel. Only inexpensive paperback edition. 255pp. 5⅜ × 8½. 25245-0 Pa. $5.95

THE TALE OF PETER RABBIT, Beatrix Potter. The inimitable Peter's terrifying adventure in Mr. McGregor's garden, with all 27 wonderful, full-color Potter illustrations. 55pp. 4¼ × 5½. (Available in U.S. only) 22827-4 Pa. $1.75

THREE PROPHETIC SCIENCE FICTION NOVELS, H. G. Wells. *When the Sleeper Wakes, A Story of the Days to Come* and *The Time Machine* (full version). 335pp. 5⅜ × 8½. (Available in U.S. only) 20605-X Pa. $5.95

APICIUS COOKERY AND DINING IN IMPERIAL ROME, edited and translated by Joseph Dommers Vehling. Oldest known cookbook in existence offers readers a clear picture of what foods Romans ate, how they prepared them, etc. 49 illustrations. 301pp. 6⅛ × 9¼. 23563-7 Pa. $6.00

SHAKESPEARE LEXICON AND QUOTATION DICTIONARY, Alexander Schmidt. Full definitions, locations, shades of meaning of every word in plays and poems. More than 50,000 exact quotations. 1,485pp. 6½ × 9¼. 22726-X, 22727-8 Pa., Two-vol. set $27.90

THE WORLD'S GREAT SPEECHES, edited by Lewis Copeland and Lawrence W. Lamm. Vast collection of 278 speeches from Greeks to 1970. Powerful and effective models; unique look at history. 842pp. 5⅜ × 8½. 20468-5 Pa. $10.95

CATALOG OF DOVER BOOKS

THE BLUE FAIRY BOOK, Andrew Lang. The first, most famous collection, with many familiar tales: Little Red Riding Hood, Aladdin and the Wonderful Lamp, Puss in Boots, Sleeping Beauty, Hansel and Gretel, Rumpelstiltskin; 37 in all. 138 illustrations. 390pp. 5⅜ × 8½. 21437-0 Pa. $5.95

THE STORY OF THE CHAMPIONS OF THE ROUND TABLE, Howard Pyle. Sir Launcelot, Sir Tristram and Sir Percival in spirited adventures of love and triumph retold in Pyle's inimitable style. 50 drawings, 31 full-page. xviii + 329pp. 6½ × 9¼. 21883-X Pa. $6.95

AUDUBON AND HIS JOURNALS, Maria Audubon. Unmatched two-volume portrait of the great artist, naturalist and author contains his journals, an excellent biography by his granddaughter, expert annotations by the noted ornithologist, Dr. Elliott Coues, and 37 superb illustrations. Total of 1,200pp. 5⅜ × 8.
Vol. I 25143-8 Pa. $8.95
Vol. II 25144-6 Pa. $8.95

GREAT DINOSAUR HUNTERS AND THEIR DISCOVERIES, Edwin H. Colbert. Fascinating, lavishly illustrated chronicle of dinosaur research, 1820's to 1960. Achievements of Cope, Marsh, Brown, Buckland, Mantell, Huxley, many others. 384pp. 5¼ × 8¼. 24701-5 Pa. $6.95

THE TASTEMAKERS, Russell Lynes. Informal, illustrated social history of American taste 1850's–1950's. First popularized categories Highbrow, Lowbrow, Middlebrow. 129 illustrations. New (1979) afterword. 384pp. 6 × 9.
23993-4 Pa. $6.95

DOUBLE CROSS PURPOSES, Ronald A. Knox. A treasure hunt in the Scottish Highlands, an old map, unidentified corpse, surprise discoveries keep reader guessing in this cleverly intricate tale of financial skullduggery. 2 black-and-white maps. 320pp. 5⅜ × 8½. (Available in U.S. only) 25032-6 Pa. $5.95

AUTHENTIC VICTORIAN DECORATION AND ORNAMENTATION IN FULL COLOR: 46 Plates from "Studies in Design," Christopher Dresser. Superb full-color lithographs reproduced from rare original portfolio of a major Victorian designer. 48pp. 9¼ × 12¼. 25083-0 Pa. $7.95

PRIMITIVE ART, Franz Boas. Remains the best text ever prepared on subject, thoroughly discussing Indian, African, Asian, Australian, and, especially, Northern American primitive art. Over 950 illustrations show ceramics, masks, totem poles, weapons, textiles, paintings, much more. 376pp. 5⅜ × 8. 20025-6 Pa. $6.95

SIDELIGHTS ON RELATIVITY, Albert Einstein. Unabridged republication of two lectures delivered by the great physicist in 1920–21. *Ether and Relativity* and *Geometry and Experience.* Elegant ideas in non-mathematical form, accessible to intelligent layman. vi + 56pp. 5⅜ × 8½. 24511-X Pa. $2.95

THE WIT AND HUMOR OF OSCAR WILDE, edited by Alvin Redman. More than 1,000 ripostes, paradoxes, wisecracks: Work is the curse of the drinking classes, I can resist everything except temptation, etc. 258pp. 5⅜ × 8½. 20602-5 Pa. $3.95

ADVENTURES WITH A MICROSCOPE, Richard Headstrom. 59 adventures with clothing fibers, protozoa, ferns and lichens, roots and leaves, much more. 142 illustrations. 232pp. 5⅜ × 8½. 23471-1 Pa. $3.95

PLANTS OF THE BIBLE, Harold N. Moldenke and Alma L. Moldenke. Standard reference to all 230 plants mentioned in Scriptures. Latin name, biblical reference, uses, modern identity, much more. Unsurpassed encyclopedic resource for scholars, botanists, nature lovers, students of Bible. Bibliography. Indexes. 123 black-and-white illustrations. 384pp. 6 × 9. 25069-5 Pa. $8.95

FAMOUS AMERICAN WOMEN: A Biographical Dictionary from Colonial Times to the Present, Robert McHenry, ed. From Pocahontas to Rosa Parks, 1,035 distinguished American women documented in separate biographical entries. Accurate, up-to-date data, numerous categories, spans 400 years. Indices. 493pp. 6½ × 9¼. 24523-3 Pa. $9.95

THE FABULOUS INTERIORS OF THE GREAT OCEAN LINERS IN HISTORIC PHOTOGRAPHS, William H. Miller, Jr. Some 200 superb photographs capture exquisite interiors of world's great "floating palaces"—1890's to 1980's: *Titanic, Ile de France, Queen Elizabeth, United States, Europa,* more. Approx. 200 black-and-white photographs. Captions. Text. Introduction. 160pp. 8⅞ × 11¾. 24756-2 Pa. $9.95

THE GREAT LUXURY LINERS, 1927–1954: A Photographic Record, William H. Miller, Jr. Nostalgic tribute to heyday of ocean liners. 186 photos of Ile de France, Normandie, Leviathan, Queen Elizabeth, United States, many others. Interior and exterior views. Introduction. Captions. 160pp. 9 × 12. 24056-8 Pa. $9.95

A NATURAL HISTORY OF THE DUCKS, John Charles Phillips. Great landmark of ornithology offers complete detailed coverage of nearly 200 species and subspecies of ducks: gadwall, sheldrake, merganser, pintail, many more. 74 full-color plates, 102 black-and-white. Bibliography. Total of 1,920pp. 8⅝ × 11¼. 25141-1, 25142-X Cloth. Two-vol. set $100.00

THE SEAWEED HANDBOOK: An Illustrated Guide to Seaweeds from North Carolina to Canada, Thomas F. Lee. Concise reference covers 78 species. Scientific and common names, habitat, distribution, more. Finding keys for easy identification. 224pp. 5⅜ × 8½. 25215-9 Pa. $5.95

THE TEN BOOKS OF ARCHITECTURE: The 1755 Leoni Edition, Leon Battista Alberti. Rare classic helped introduce the glories of ancient architecture to the Renaissance. 68 black-and-white plates. 336pp. 8⅜ × 11¼. 25239-6 Pa. $14.95

MISS MACKENZIE, Anthony Trollope. Minor masterpieces by Victorian master unmasks many truths about life in 19th-century England. First inexpensive edition in years. 392pp. 5⅜ × 8½. 25201-9 Pa. $7.95

THE RIME OF THE ANCIENT MARINER, Gustave Doré, Samuel Taylor Coleridge. Dramatic engravings considered by many to be his greatest work. The terrifying space of the open sea, the storms and whirlpools of an unknown ocean, the ice of Antarctica, more—all rendered in a powerful, chilling manner. Full text. 38 plates. 77pp. 9¼ × 12. 22305-1 Pa. $4.95

THE EXPEDITIONS OF ZEBULON MONTGOMERY PIKE, Zebulon Montgomery Pike. Fascinating first-hand accounts (1805–6) of exploration of Mississippi River, Indian wars, capture by Spanish dragoons, much more. 1,088pp. 5⅜ × 8½. 25254-X, 25255-8 Pa. Two-vol. set $23.90

A CONCISE HISTORY OF PHOTOGRAPHY: Third Revised Edition, Helmut Gernsheim. Best one-volume history—camera obscura, photochemistry, daguerreotypes, evolution of cameras, film, more. Also artistic aspects—landscape, portraits, fine art, etc. 281 black-and-white photographs. 26 in color. 176pp. 8⅜ × 11¼. 25128-4 Pa. $12.95

THE DORÉ BIBLE ILLUSTRATIONS, Gustave Doré. 241 detailed plates from the Bible: the Creation scenes, Adam and Eve, Flood, Babylon, battle sequences, life of Jesus, etc. Each plate is accompanied by the verses from the King James version of the Bible. 241pp. 9 × 12. 23004-X Pa. $8.95

HUGGER-MUGGER IN THE LOUVRE, Elliot Paul. Second Homer Evans mystery-comedy. Theft at the Louvre involves sleuth in hilarious, madcap caper. "A knockout."—Books. 336pp. 5⅜ × 8½. 25185-3 Pa. $5.95

FLATLAND, E. A. Abbott. Intriguing and enormously popular science-fiction classic explores the complexities of trying to survive as a two-dimensional being in a three-dimensional world. Amusingly illustrated by the author. 16 illustrations. 103pp. 5⅜ × 8½. 20001-9 Pa. $2.00

THE HISTORY OF THE LEWIS AND CLARK EXPEDITION, Meriwether Lewis and William Clark, edited by Elliott Coues. Classic edition of Lewis and Clark's day-by-day journals that later became the basis for U.S. claims to Oregon and the West. Accurate and invaluable geographical, botanical, biological, meteorological and anthropological material. Total of 1,508pp. 5⅜ × 8½.
21268-8, 21269-6, 21270-X Pa. Three-vol. set $25.50

LANGUAGE, TRUTH AND LOGIC, Alfred J. Ayer. Famous, clear introduction to Vienna, Cambridge schools of Logical Positivism. Role of philosophy, elimination of metaphysics, nature of analysis, etc. 160pp. 5⅜ × 8½. (Available in U.S. and Canada only) 20010-8 Pa. $2.95

MATHEMATICS FOR THE NONMATHEMATICIAN, Morris Kline. Detailed, college-level treatment of mathematics in cultural and historical context, with numerous exercises. For liberal arts students. Preface. Recommended Reading Lists. Tables. Index. Numerous black-and-white figures. xvi + 641pp. 5⅜ × 8½.
24823-2 Pa. $11.95

28 SCIENCE FICTION STORIES, H. G. Wells. Novels, *Star Begotten* and *Men Like Gods,* plus 26 short stories: "Empire of the Ants," "A Story of the Stone Age," "The Stolen Bacillus," "In the Abyss," etc. 915pp. 5⅜ × 8½. (Available in U.S. only)
20265-8 Cloth. $10.95

HANDBOOK OF PICTORIAL SYMBOLS, Rudolph Modley. 3,250 signs and symbols, many systems in full; official or heavy commercial use. Arranged by subject. Most in Pictorial Archive series. 143pp. 8⅛ × 11. 23357-X Pa. $5.95

INCIDENTS OF TRAVEL IN YUCATAN, John L. Stephens. Classic (1843) exploration of jungles of Yucatan, looking for evidences of Maya civilization. Travel adventures, Mexican and Indian culture, etc. Total of 669pp. 5⅜ × 8½.
20926-1, 20927-X Pa., Two-vol. set $9.90

AMERICAN CLIPPER SHIPS: 1833–1858, Octavius T. Howe & Frederick C. Matthews. Fully-illustrated, encyclopedic review of 352 clipper ships from the period of America's greatest maritime supremacy. Introduction. 109 halftones. 5 black-and-white line illustrations. Index. Total of 928pp. 5⅜ × 8½.
25115-2, 25116-0 Pa., Two-vol. set $17.90

TOWARDS A NEW ARCHITECTURE, Le Corbusier. Pioneering manifesto by great architect, near legendary founder of "International School." Technical and aesthetic theories, views on industry, economics, relation of form to function, "mass-production spirit," much more. Profusely illustrated. Unabridged translation of 13th French edition. Introduction by Frederick Etchells. 320pp. 6⅛ × 9¼. (Available in U.S. only) 25023-7 Pa. $8.95

THE BOOK OF KELLS, edited by Blanche Cirker. Inexpensive collection of 32 full-color, full-page plates from the greatest illuminated manuscript of the Middle Ages, painstakingly reproduced from rare facsimile edition. Publisher's Note. Captions. 32pp. 9⅜ × 12¼. 24345-1 Pa. $4.50

BEST SCIENCE FICTION STORIES OF H. G. WELLS, H. G. Wells. Full novel *The Invisible Man,* plus 17 short stories: "The Crystal Egg," "Aepyornis Island," "The Strange Orchid," etc. 303pp. 5⅜ × 8½. (Available in U.S. only)
21531-8 Pa. $4.95

AMERICAN SAILING SHIPS: Their Plans and History, Charles G. Davis. Photos, construction details of schooners, frigates, clippers, other sailcraft of 18th to early 20th centuries—plus entertaining discourse on design, rigging, nautical lore, much more. 137 black-and-white illustrations. 240pp. 6⅛ × 9¼.
24658-2 Pa. $5.95

ENTERTAINING MATHEMATICAL PUZZLES, Martin Gardner. Selection of author's favorite conundrums involving arithmetic, money, speed, etc., with lively commentary. Complete solutions. 112pp. 5⅜ × 8½. 25211-6 Pa. $2.95

THE WILL TO BELIEVE, HUMAN IMMORTALITY, William James. Two books bound together. Effect of irrational on logical, and arguments for human immortality. 402pp. 5⅜ × 8½. 20291-7 Pa. $7.50

THE HAUNTED MONASTERY and THE CHINESE MAZE MURDERS, Robert Van Gulik. 2 full novels by Van Gulik continue adventures of Judge Dee and his companions. An evil Taoist monastery, seemingly supernatural events; overgrown topiary maze that hides strange crimes. Set in 7th-century China. 27 illustrations. 328pp. 5⅜ × 8½. 23502-5 Pa. $5.00

CELEBRATED CASES OF JUDGE DEE (DEE GOONG AN), translated by Robert Van Gulik. Authentic 18th-century Chinese detective novel; Dee and associates solve three interlocked cases. Led to Van Gulik's own stories with same characters. Extensive introduction. 9 illustrations. 237pp. 5⅜ × 8½.
23337-5 Pa. $4.95

Prices subject to change without notice.

Available at your book dealer or write for free catalog to Dept. GI, Dover Publications, Inc., 31 East 2nd St., Mineola, N.Y. 11501. Dover publishes more than 175 books each year on science, elementary and advanced mathematics, biology, music, art, literary history, social sciences and other areas.